Peter Demian
Renate Fruchter

CoMem: Design Knowledge Reuse from a Corporate Memory

Peter Demian
Renate Fruchter

CoMem: Design Knowledge Reuse from a Corporate Memory

How to Find and Understand Designs from
Previous Design Projects in a Corporate
Repository

VDM Verlag Dr. Müller

Imprint

Bibliographic information by the German National Library: The German National Library lists this publication at the German National Bibliography; detailed bibliographic information is available on the Internet at http://dnb.d-nb.de.

Cover image: www.purestockx.com

Publisher:
VDM Verlag Dr. Müller Aktiengesellschaft & Co. KG, Dudweiler Landstr. 125 a,
66123 Saarbrücken, Germany,
Phone +49 681 9100-698, Fax +49 681 9100-988,
Email: info@vdm-verlag.de

Produced in USA and UK by:
Lightning Source Inc., La Vergne, Tennessee, USA
Lightning Source UK Ltd., Milton Keynes, UK
BookSurge LLC, 5341 Dorchester Road, Suite 16, North Charleston, SC 29418, USA

ISBN: 978-3-639-05664-8

ABSTRACT

The objective of this research is to improve and support the process of design knowledge reuse in the architecture, engineering, and construction industry. Whereas internal knowledge reuse (reusing from one's personal memory or experiences) is very effective, external knowledge reuse (reusing from an external digital or paper archive) often fails. Ethnographic observations show that the three key activities in the internal knowledge reuse process are:

- *Finding* a reusable item
- Exploring this item's project context which leads to *understanding*
- Exploring this item's evolution history which leads to *understanding*

The approach of this research is to design and support the external reuse process so that it matches the internal reuse process. The hypothesis is that if the designer's interaction with the external repository enables him/her to:

- Rapidly *find* relevant items of design knowledge
- View each item in context in order to *understand* it, specifically:
  - Explore its project context
  - Explore its evolution history

then the process of reuse will be improved. This research addresses the following questions: (i) how do finding and understanding occur in internal knowledge reuse, and (ii) how can they be supported in external reuse?

The internal knowledge reuse aspects of these questions are formalized based on an ethnographic study. The findings of the study indicate that *finding* is effective in internal knowledge reuse because the designer has a succinct overview of the entire corporate memory in his/her head, and can gauge each item's relevance to the current design task. *Understanding* occurs in internal knowledge reuse by exploring the context of the item being reused. Two contextual dimensions are typically explored: (i) the project context, and (ii) the evolution history. This research presents six degrees of exploration that lead to understanding.

Internal knowledge reuse therefore consists of three steps: finding a potentially reusable item, exploring that item's project context, and exploring that item's evolution history. This research presents a prototype system, CoMem (Corporate Memory), which consists of three modules, one to support each of these steps.

The CoMem Overview explores how finding reusable design knowledge may be supported in external repositories. It is implemented in the form of a Corporate Map that presents a succinct snapshot of the entire corporate memory that enables the user to make multi-granularity comparisons and quickly find reusable items. One innovative aspect of the Corporate Map is that each item is color-coded by its relevance to the user's design task. The relevance measure is the result of applying information retrieval techniques to the corpus of corporate memory design objects based on a query representing the current design task. An in-depth study of how this relevance may be measured is presented.

i

Given an item from the map that the designer is considering reusing, the CoMem Project Context Explorer identifies related items in the corporate memory, and visually presents these related contextual items to help the user better understand why the item in question was designed the way it was. It identifies related items by combining the CoMem relevance measure with the classic fisheye formulation.

Storytelling is one of the best knowledge transfer mechanisms. The CoMem Evolution History Explorer presents the multiple versions of the item in question, and the team interactions and rationale driving this evolution. It draws from the effectiveness of comic books for telling stories, and explores how this effectiveness can be carried over to the presentation of version histories.

Finally, a usability evaluation of the CoMem prototype is performed using formal user testing. For this purpose, a usability testing framework and methodology is proposed. The key dimensions for the usability testing are the size of the repository, and the type of finding task: exploration versus retrieval. This research highlights the importance of exploration, which is normally overlooked by traditional tools.

The evaluation results show that CoMem offers greater support for finding and understanding than traditional tools, and reuse using CoMem is consistently rated to be more effective by test participants. This supports the hypothesis that finding and understanding lead to more effective reuse. This research makes important contributions by formalizing the reuse process, developing an innovative tool to support that process, and building a framework to study and assess such tools.

ACKNOWLEDGMENTS

I first wish to thank God, my parents Dr. Samir Demian and Dr. Fadia Mikhail, and my siblings Caroline and Mark. The research presented here was completed during a PhD at Stanford University. I am deeply grateful to my principal adviser and co-author, Dr. Renate Fruchter. Over the course of my studies at Stanford I have come to realize that she is a wonderful educator and mentor, and quite simply a great person. She was of immense support to me and my family during and after my injury. I am also grateful to my other PhD committee-members, Professor Helmut Krawinkler, Professor Terry Winograd, and Professor Kincho Law. I am indebted to Professor Prabhakar Raghavan, who at the time was Vice-President and Chief Technology Officer of Verity, Inc. and consulting professor at Stanford, for his guidance on the CoMem relevance measure. He had several informal discussions with me and it was he who first suggested to me the tree isomorphism method.

CoMem uses the Treemap Library by Christophe Bouthier, LSI by Telcordia Technologies, and the Jazz Library by HCIL at the University of Maryland.

Dr. Greg Luth, founder and president of GPLA, is a long-time friend of the research team, and provided much needed real-world input into this research, for which I thank him.

My Master's degree at Stanford was funded by a fellowship from the School of Engineering, and the Doctorate by the UPS Endowment at Stanford, the Center for Integrated Facility Engineering, and the Project-Based Learning Laboratory.

Finally, I thank the Coptic community in Northern California between 1998 and 2003 and all who were like a family to me during my time there. Any attempt to name them all would be at once unreadable and scandalously incomplete, and so I simply ask God to reward them for their love.

Peter Demian
Originally written in Stanford, June 2004
Revised in Loughborough, June 2008

TABLE OF CONTENTS

vi

LIST OF TABLES

LIST OF FIGURES

Chapter 1

INTRODUCTION

The average designer, whether consciously or subconsciously, draws from a vast well of previous design experience. "All design is redesign" (Leifer 1997). This can be experience acquired by the individual or by his/her mentors or professional community. This activity is referred to as *design knowledge reuse*. Specifically, this research defines design knowledge reuse as the reuse of previously designed artifacts or components, as well as the knowledge and expertise ingrained in these previous designs. This research distinguishes between two types of reuse:

1. *Internal knowledge reuse*: a designer reusing knowledge from his/her own personal experiences (internal memory).
2. *External knowledge reuse*: a designer reusing knowledge from an external knowledge repository (external memory).

Internal knowledge reuse is a very effective process, which some writers place at the very center of human intelligence:

> *We get reminded of what has happened to us previously for a very good reason. Reminding is the mind's method of coordinating past events with current events to enable generalization and prediction. Intelligence depends upon the ability to translate descriptions of new events into labels that help in the retrieval of prior events. One can't be said to know something if one can't find it in memory when it is needed. Finding a relevant past experience that will help make sense of a new experience is at the core of intelligent behavior. (Schank 1990, pages 1, 2)*

On the other hand external knowledge reuse often fails. This failure occurs for numerous reasons, including:

- To be available for external reuse, knowledge needs to be *captured* and *stored* in an external repository. Designers do not appreciate the importance of knowledge capture because of the additional overhead required to document their process and rationale. They perceive that capture and reuse costs more than recreation from scratch. Consequently, knowledge is often not captured.
- Even when knowledge capture does take place, it is limited to formal knowledge (e.g. documents). Contextual or informal knowledge, such as the rationale behind design decisions, or the interaction between team members on a design team, is often lost, rendering the captured knowledge not reusable, as is often the case in current industry documentation practices.
- There are no mechanisms from both the information technology and organizational viewpoints for finding and retrieving reusable knowledge or exploring external repositories.

Our empirical observations of designers at work show that internal knowledge reuse is effective since:

- The designer can quickly *find* (mentally) reusable items.
- The designer can remember the context of each item, and can therefore *understand* it and reuse it more effectively.

These observations of *internal knowledge reuse* are used as the basis for improving *external knowledge reuse*.

Knowledge reuse is viewed as a step in the knowledge life cycle (Figure 1). Knowledge is created as designers collaborate on design projects. It is captured, indexed, and stored in an archive. At a later time, it is retrieved from the archive and reused. Finally, as knowledge is reused, it is refined and becomes more valuable. In this sense, the archive system acts as a knowledge refinery. This research focuses on the knowledge reuse phase and builds on previous work that addresses knowledge creation, capture, indexing, and archiving (Fruchter 1996, Fruchter et al. 1998, Reiner and Fruchter 2000).

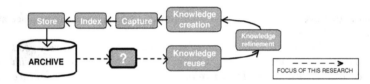

Figure 1: The knowledge life cycle. Knowledge is created, captured, indexed, and stored in an archive. At a later time, it is retrieved from the archive and reused. As it is reused, it becomes refined. This research focuses on the knowledge reuse phase.

Practical Motivation

The motivation behind the development of external knowledge reuse systems is that the capture and reuse of knowledge is less costly than its recreation. In many architecture, engineering, and construction (AEC) firms today, knowledge capture and reuse is limited to dealing with paper archives. Even when the archives are digital, they are usually in the form of electronic files (documents) arranged in folders which are difficult to explore and navigate. A typical query might be, "how did we design previous cooling tower support structures in hotel building projects?" In many cases, the user of such systems is overloaded with information, but with very little context to help him/her decide if, what, and how to reuse.

This research addresses the following central questions:

- What are the key characteristics of the *internal* knowledge reuse process, and how can a similar process be supported in the case of *external* knowledge reuse?
- What are natural idioms that can be modeled into a computer system to provide an effective knowledge reuse experience to a designer?

2

Scope and Assumptions

This research aims to *support* rather than *automate* the process of design knowledge reuse. By observing how internal knowledge reuse occurs naturally in practicing designers, this research develops interaction metaphors and retrieval mechanisms that compliment and assist this natural knowledge reuse process.

This research concentrates on *design* knowledge reuse, i.e. *actual designs* and *project content* produced by designers working on design projects. The term "design knowledge reuse" is used rather than "design reuse" to indicate that what is reused is often more than just previously designed artifacts, but also includes the knowledge and expertise ingrained in these previous designs. Specifically, this research uses the term "design knowledge" to refer to design knowledge as it is captured by the *Semantic Modeling Engine* (SME) (Fruchter 1996).

There are two possible lines of attack for addressing the problem of reuse from an external repository (Figure 2):

- **Retrieval approach.** The repository is treated as a corpus of documents. The user has an information need, which he/she translates into a query. The system takes this query as its input and returns a set of (ranked) items as its output.
- **Exploration approach.** The user *explores* the repository. This process of exploration is equally as important as the items that are eventually retrieved in satisfying the user's information need.

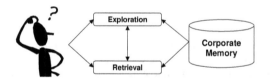

Figure 2: Two approaches for supporting reuse from an external repository[1].

The choice of interaction method (exploration versus retrieval) will affect the nature of the content that needs to be stored in the corporate memory. This research will focus on exploration rather than retrieval, although retrieval is revisited during the evaluation phase. The user will interact with rich, detailed, interlinked content rather than a collection of simple documents in the corporate memory. The problem of supporting design knowledge reuse will be framed in terms of creating interaction experiences in which the external repository can be visualized and explored. This approach is based on the following assumptions:

---

[1] This figure is adapted from Baeza-Yates and Ribeieo-Neto 1999.

- Humans are highly attuned to images and visual information. Visual representations communicate certain kinds of information more effectively than any other methods (Card et al. 1999).
- We are dealing with large amounts of design knowledge and so exploration might be more effective than retrieval.
- Retrieval techniques will not be effective because this design knowledge is not formally represented (as a result of the extra effort required of the designer to capture design rationale formally).

This research was carried out in the context of the AEC industry, and particularly within the field of structural engineering. A knowledge reuse model was developed based on scenarios of structural designers working on construction projects.

Although the emphasis is on structural engineering, the interaction of the structural subsystem with other building subsystems is considered to be a key element in the notion of knowledge in context.

Issues of liability and ownership of knowledge, although undoubtedly extremely important, are beyond the scope of this study.

The Importance of the Knowledge Reuse Problem

Design Perspective
Why is design knowledge reuse an important issue? From a design perspective, the crucial concern is the tradeoff between productivity and creativity. At one extreme, the designer can choose not to reuse any knowledge at all from prior work. If successful, this approach can lead to an extremely creative solution; if unsuccessful, it can waste a lot of time, with very little added value in the quality of the solution ("reinventing the wheel"). The second extreme is for the designer to reuse a lot of knowledge (or even an entire solution) from the well of previous design experience. If successful, this approach can save resources and lead to a better solution (for example, a novice learning from previous solutions created by experts); if unsuccessful, this approach can result in previous knowledge being reused inappropriately.

It is therefore important for the designer to take an approach which is somewhere in between the two extremes. In exploring this middle ground, the designer needs to ask questions such as:
- *Can I reuse anything from past experiences?* Are there similar situations captured in the external repository that might be useful?
- *How much can I reuse?* Small details or large portions of the design?
- *What should I reuse?* Actual physical components? Abstract concepts or ideas? Lessons learned from previous design processes? Design tools or analysis tools?

4

The underlying principle is that reuse should save resources (time and money), but not at the expense of the quality of the final design.

Business Perspective
From a business perspective, an effective knowledge reuse strategy needs to enable a corporation to retain and reuse the knowledge accumulated from many years of experience. Specifically it should:
- Reduce the time wasted on recreating knowledge.
- Reduce the time wasted on searching for knowledge in obsolete archives.
- Retain knowledge in the corporation even after the retirement or departure of knowledgeable employees.

A knowledge reuse system can also be thought of as a learning resource:
- Novices can learn and benefit from the expertise of more experienced employees.
- Best practices are captured and reused by employees.

The underlying principle is that knowledge is a company's most important strategic resource, which, if properly managed, can drastically improve the company's productivity and lead to a greater competitive advantage.

Research Hypothesis
The objective of this research is to *improve* and *support* the process of design knowledge reuse in the AEC industry. Based on observations of internal knowledge reuse from an ethnographic study, the three key activities in the knowledge reuse process are:
- *Finding* a reusable item.
- Exploring this item's *project context* which leads to *understanding*.
- Exploring this item's *evolution history* which leads to *understanding*.

Hypothesis:
If the designer's interaction with the external repository enables him/her to:
- Rapidly *find* relevant items of design knowledge.
- View each item *in context* in order to *understand* its appropriateness, specifically:
  - Explore its project context.
  - Explore its evolution history.
- ⇨ Then the process of reuse will be improved.

This improved reuse will lead to higher quality design solutions, and save time and money.

Research Questions
This research addresses the following questions:

Question 1: How does *finding* occur in internal knowledge reuse? What retrieval mechanisms are needed to support the *finding* of reusable design knowledge in a large corporate repository of design content? What are suitable interaction metaphors and visualization techniques?

Question 2: What is the nature of the project context exploration in internal knowledge reuse? How can this exploration be supported in a large corporate repository of design content? What are suitable interaction metaphors and visualization techniques?

Question 3: What is the nature of the evolution history exploration in internal knowledge reuse? How can this exploration be supported in a large corporate repository of design content? What are suitable interaction metaphors and visualization techniques?

Points of Departure
Design as reflection-in-action. This research is the latest in a line of research projects on design knowledge management conducted at the Project-Based Learning Lab at Stanford University. These projects are based on Schön's *reflective practitioner* paradigm of design (Schön 1983). Schön argues that every design task is unique, and that the basic problem for designers is to determine how to approach such a single unique task. Schön places this tackling of unique tasks at the center of design practice, a notion he terms *knowing-in-action*:

> *Once we put aside the model of Technical Rationality which leads us to think of intelligent practice as an application of knowledge... there is nothing strange about the idea that a kind of knowing is inherent in intelligent action... it does not stretch common sense very much to say that the know-how is in the action – that a tight-rope walker's know-how, for example, lies in and is revealed by, the way he takes his trip across the wire... There is nothing in common sense to make us say that the know-how consists in rules or plans which we entertain in the mind prior to action. (Schön 1983, page 50)*

To Schön, design, like tightrope walking, is an *action-oriented* activity. However, when knowing-in-action breaks down, the designer may consciously transition to acts of reflection. Schön calls this *reflection-in-action*. In a cycle which Schön refers to as a *reflective conversation with the situation*, designers reflect by *naming* the relevant factors, *framing* the problem in a certain way, making *moves* toward a solution and *evaluating* those moves. Schön argues that, whereas action-oriented knowledge is often tacit and difficult to express or convey, what *can* be captured is reflection-in-action.

Semantic Modeling Engine. This reflection-in-action cycle forms the conceptual basis of knowledge capture in the *Semantic Modeling Engine* (SME) (Fruchter 1996). SME is a framework that enables designers to map objects from a shared product model to

6

multiple semantic representations and to other shared project knowledge. Figure 3 shows a simplified entity-relationship diagram of the SME schema (Figure 3 (a)), and an example of actual project knowledge (Figure 3(b)). In SME, a *project object* encapsulates multiple *discipline objects*, and a *discipline object* encapsulates multiple *component objects*. Each SME object can be linked to graphic objects from the shared 3D product model, or to other shared project documents or data (such as vendor information, analysis models, sketches, calculations).

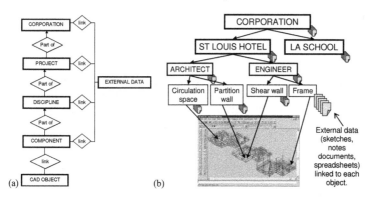

Figure 3: (a) A simplified entity-relationship diagram of the SME schema. A *project object* contains multiple *discipline objects*, and a *discipline object* contains multiple *component objects*. These semantic objects can be linked to graphic objects or to external data. (b) An example of actual project knowledge in SME.

SME supports Schön's reflection-in-action by enabling the designer to declare his/her particular perspective on the design (i.e. *framing* the problem) by creating a discipline object. Next he/she proceeds to *name* the individual components of the problem as he/she sees it by creating component objects. SME discipline objects are exported to external analysis tools to derive building behavior and evaluate it by comparing it to functional requirements (Eastman 1999). The designer uses these as the basis for making design decisions, i.e., making *moves* towards the solution and *evaluating* those moves.

Project Memory. The ProMem (Project Memory) system (Fruchter et al. 1998, Reiner and Fruchter 2000) takes the Semantic Modeling Engine as its point of departure and adds to it the time dimension. ProMem captures the evolution of the project at the three levels of granularity identified by SME as emulating the structure of project knowledge: *project*, *discipline*, and *component*. ProMem automatically versions each SME object every time a change is made in the design or additional knowledge is created. This versioning is transparent to the designer. The designer is able to go back and flag any version to indicate its *level of importance* (*low, conflict*, or *milestone*) and its *level of sharing* (*private, public*, or *consensus*).

7

Corporate Memory. This research presents CoMem (Corporate Memory), a prototype system that extends ProMem firstly by grouping the accumulated set of project memories into a *corporate memory*, and secondly by supporting the designer in reusing design knowledge from this corporate memory in new design projects. This support for knowledge reuse is based on our observations of internal knowledge reuse by designers at work. This knowledge reuse is not limited to designed components and subcomponents, but includes the evolution, rationale, and domain expertise that contributed to these designs. Here this research echoes Schön's contention that design expertise lies not in "rules or plans entertained in the mind prior to action" but in the action itself.

Monograph Roadmap

Chapter 2 presents an overview of related research centered on the three themes of *knowledge*, *design*, and *reuse*. It also introduces ideas from the fields of information retrieval and human-computer interaction that are considered points of departure for this research.

Chapter 3 presents findings from an ethnographic study of knowledge reuse amongst AEC practitioners. The results from this study offer insights into the process of internal knowledge reuse and have important implications for the design of a computer system for supporting external knowledge reuse.

Chapter 4 describes the methodology for this research. The results from the ethnographic study are distilled into a few main points. The chapter describes how these ethnographic findings were used to design CoMem using a scenario-based method, and how CoMem was evaluated.

Chapter 5 gives an introduction to the three CoMem modules, the *Overview*, the *Project Context Explorer*, and the *Evolution History Explorer*. Chapter 6, Chapter 7, and Chapter 8 consider each module separately. In particular, these chapters describe the task that each module is intended to support, and explore how those tasks may be supported using visualization techniques and interaction metaphors. Chapter 9 summarizes the CoMem modules through the use of a typical CoMem usage scenario.

Chapter 10 examines the problem of measuring relevance in CoMem. Relevance measurements are used in both the Overview and the Project Context Explorer. The chapter presents and evaluates several techniques for measuring relevance based on text analysis and introduces an innovative technique for analyzing hierarchical data based on the problem of tree isomorphism.

Chapter 11 presents a formal evaluation of CoMem as a whole. CoMem is compared to more traditional tools for different types of tasks (retrieval versus exploration) and repository sizes (large versus small).

This monograph concludes with Chapter 12, which presents a discussion of the research results and contributions in light of the stated hypothesis and research questions, and closes with the conclusions that can be drawn from this research. Finally, the road is paved for future research including a usability framework for designing and analyzing information interfaces based on the three dimensions of type of tasks (retrieval versus exploration), repository sizes (large versus small), and levels of familiarity (familiar versus unfamiliar).

Chapter 2

RELATED RESEARCH

The concept of *design knowledge reuse* is at the intersection of three other concepts: *knowledge, design,* and *reuse* (Figure 4).

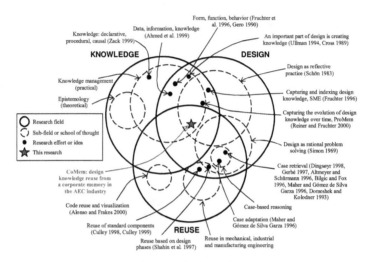

Figure 4: Related research.

Knowledge

Knowledge has been studied in a wide variety of contexts. At the theoretical end of the spectrum is the field of epistemology, which is concerned with the nature of knowledge, especially its limits and validity (Pollock and Cruz 1999 gives an overview). Epistemological insights into the role of the human memory can guide the process of designing an external memory system.

Memory plays a central role in *inductive reasoning*. In enumerative induction, one examines a sample of objects of some kind, A, observes that all the As in the sample have another property B, and infers on that basis that all As are Bs. Statistical induction is a variation wherein one observes that some proportion m/n of As in the sample are Bs, and then infers that the probability of an arbitrary A being a B is m/n.

Memory supplies us with premises for arguments. These premises are typically themselves the conclusions of earlier arguments, but these earlier arguments do not have to be rehearsed in order to make use of their conclusions. In other words, people remember conclusions but not reasons. In this way memory acts as a source of knowledge.

However memory is more than just a source of premises. Memory also supplies us with *defeaters*, reasons for rejecting a previously held belief. This has implications for the corporate memory as a knowledge refinery. Epistemologists note that, while memory search is not conscious, it is more than just searching through facts. We are somehow always on the lookout for newly inferred defeaters for previous steps of reasoning.

The more practical field of *knowledge management* is more closely related to this research, although some researchers (for example von Krogh et al. 1998) have sought to reassess the knowledge management research agenda by appealing to profound epistemological theories.

In the knowledge management literature, knowledge is commonly distinguished from data and information. Broadly speaking, data are observations or facts out of context; information results from placing data within a meaningful context. Knowledge is "that which we come to believe and value based on the meaningfully organized accumulation of information through experience, communication or inference" (Zack 1999).

It has been rightfully noted that data, information, and knowledge are relative concepts (Ahmed et al. 1999). Although the precise distinctions between the three are not of immediate interest, there is clearly some dimension along which data would be ranked near the bottom and knowledge near the top. Intuitively, this dimension is closely related to *context*. Context is the framework within which information can be interpreted and understood. To clarify this notion of context, two commonly used knowledge classifications are presented below.

Declarative, procedural, causal. In this research, the term "*design* knowledge" is taken to refer to knowledge *about* a certain artifact (*declarative* knowledge), for example the dimensions of a cooling tower frame. However, if a designer were to reuse this cooling tower frame in a new project, he/she would need to know *how* the original dimensions were calculated (*procedural* knowledge), and *why* they were given those values (*causal* knowledge).

Form, function, behavior. Within the field of design theory and methodology, knowledge related to an artifact is often categorized into *form* (or *structure*), *function*, and *behavior* (Gero 1990). An artifact's form is knowledge about its physical composition; its function is knowledge about what it should do; and its behavior is knowledge about what it actually does, or how well it performs.

Declarative knowledge is the principle output of the design process, but it is rendered more reusable if it is enriched with procedural and causal knowledge. Similarly,

11

knowledge of the function and behavior is a useful supplement to knowledge of the form.

This research does not propose to make use of formal knowledge classifications. The important point to make is that the knowledge that is typically considered to be the output of the design process (i.e. the description of an artifact which enables someone to build it) is usually lacking the context which would enable this knowledge to be reused in the future. This is what is meant by *knowledge in context*; i.e. the additional knowledge that is generated or used during the design process, but which is not traditionally communicated as the output of the design process.

In order for knowledge to be reusable, it has to be as rich as possible, i.e. it has to be presented in the context in which it was created. This requirement may pose many challenges for knowledge capture because contextual knowledge is often *tacit* (Polanyi 1966), i.e. not encoded at all, or embedded in informal media, or impossible to detach from the people processing it. For example, Brown and Duguid (2000) write:

> *Knowledge entails a knower. Where people treat information as independent and more-or-less self-sufficient, they seem more inclined to associate knowledge with someone. In general, it seems right to ask, "Where is that information?" but odd to ask, "Where's that knowledge?"... It seems more reasonable to ask, "Who knows that?" Second, given this personal attachment, knowledge appears harder to detach than information. People treat information as self-contained substance. It is something that people pick up, possess, pass around, put in a database, lose, find, write down, accumulate, count, compare, and so forth. Knowledge, by contrast, doesn't take as kindly to ideas of shipping, receiving, and quantification. It is hard to pick up and hard to transfer... Third, one reason why knowledge may be so hard to give and receive is that knowledge seems to require more by way of assimilation. Knowledge is something we digest rather than merely hold. It entails the knower's understanding and some degree of commitment... while it seems quite reasonable to say, "I've got the information, but I don't understand it," it seems less reasonable to say, "I know, but I don't understand." (Brown and Duguid 2000, pages 119-120)*

Design
There exist several definitions of design, as well as design process models, design theories, and design methodologies (Dorst 1997, Cross 1989). In the context of design knowledge reuse, the following definition by Ullman seems fitting: "design is the process of developing information about an object that has not previously existed" (Ullman 1994). Cross (1989) makes a similar assertion: "the most essential design activity is the production of a final description of the artifact."

12

These statements about design are useful because they emphasize that design is an activity that generates knowledge, and implicitly this knowledge can be reused. However, they sidestep the crucial issue of how designs are generated, that is the "creative" part of design. It has been argued above that simply capturing the knowledge that is produced at the end of the design process is not enough. Supplementing descriptive knowledge about an artifact with contextual knowledge requires some understanding of the "inner workings" of the design process. The *process* by which the designed artifact evolves needs to be captured.

This "black box" of creative design has been the subject of much research. The earliest design researchers viewed design as a rational (or rationalizable) process made up of distinct phases. Later, attempts to incorporate more theoretical knowledge of designers and design problems into these rational phase models led to the view of design as *rational problem solving* (Simon 1969). Later still, perhaps as a reaction, fundamentally different views emerged, which took a phenomenological approach and regarded design as a subjective and deeply human experience (Schön 1983). Which paradigm best describes the design process as experienced by designers is an ongoing line of research (Dorst 1997).

The question of design paradigms is not central to this research. As noted above, this research continues along the path set by the SME (Fruchter 1996) and ProMem (Reiner and Fruchter 2000) research projects. These projects are based on the *Reflective Practitioner* paradigm (Schön 1983).

Design Reuse
Although much research is dedicated to design theory and design knowledge capture, considerably less focuses specifically on reuse. Research studies on design knowledge reuse focus either on the *cognitive* aspects or on the *computational* aspects.

Research into the cognitive aspects of reuse has helped to identify the information needed by designers. Kuffner and Ullman (1990) found that the majority of information requested by mechanical engineers was concerning the operation or purpose of a design object, information that is not typically captured in standard design documents (drawings and specifications). Finger (1998) observed that designers rarely use CAD tools to help them organize and retrieve design information. This research extends these findings by formalizing the requirements for contextual information when reusing items from previous projects. Ye and Fischer (2002) go further, noting that an important cognitive barrier to external reuse is the user's unfamiliarity with the contents of the repository. Users are not aware of what is in the repository and so do not know to look for it. They cannot anticipate the existence of a reusable component in the repository.

On the computational side, research into design knowledge reuse focuses on design knowledge *representation* and *reasoning*. Knowledge representation ranges from

13

informal classification systems for standard components[2] (see for example Culley 1998, Culley 1999) to more structured design rationale approaches (Regli et al. 2000 gives an overview). There is a tradeoff in design rationale systems between the overhead for recording design activities and the structure of the knowledge captured. History-based rationale approaches, such as electronic notebooks (Lakin et al. 1989), require a low overhead but result in a collection of disparate documents. Argumentation-based approaches (McCall 1987, Chung and Goodwin 1994) and device-based approaches (Baudin et al. 1993) provide a more uniform structure, but add a documentation overhead to the design process.

Highly structured representations of design knowledge can be used for *reasoning*. Two common approaches are case-based reasoning and model-based reasoning. However, these approaches usually require manual pre or post processing, structuring and indexing of design knowledge.

This research brings together the cognitive and computational approaches. It considers reuse to be a combined effort involving both the human and the computer. Therefore the issue of design knowledge reuse is addressed as a human-computer interaction problem, and a user-centered approach is taken to designing this interaction. The aim is to provide a knowledge reuse experience that leverages natural idioms and metaphors in order to support the designer in doing his/her work, and automatic reasoning approaches are considered to constrain the user's knowledge reuse activities. In this approach, capture and indexing take place in real time, with the least possible intrusion on the design process. Knowledge is captured by supporting the typical communication and coordination activities that occur during collaborative design.

Three research areas related to the computational aspects of design reuse deserve special attention:
- Case-based and model-based reasoning (AEC industry)
- Reuse models (mostly mechanical engineering)
- Code reuse (software engineering)

Case-Based Reasoning, Case-Based Design, Model-Based Reasoning
The principle that "all design is redesign" expresses the idea that designers are inevitably influenced by things that they or others have designed in the past. The term "redesign" implies that new designs can be created by modifying old designs. This is the premise behind using *case-based reasoning* to automate some aspects of the design process.

The differences between this research and case-based reasoning are summarized in Table 1.

---

[2] It has been argued that component reuse should not be restricted to standard parts coming from catalogs but should also include reuse of designed components (Culley and Theobald 1997).

14

Table 1: Differences between this research and case-based reasoning.

	This research	Case-based reasoning
FUNDAMENTAL DIFFERENCES:		
Design is…	Collaborative reflection	Rational problem solving
Research objective is…	To support the design process	To automate the design process[3]
CONSEQUENCES:		
Knowledge representation:	Informal, facilitate collaboration	Formal, *a priori* schema
Role of human:	To do design (evolution captured transparently)	To input previous design cases (high overhead)
Reuse mechanism:	Human designer explores corporate memory – knowledge in context	Automated reasoning based on previous cases

ARCHIE is a case-based reasoning tool for aiding architects during conceptual design (Domeshek and Kolodner 1993). ARCHIE breaks down previous design cases into "chunks", and uses indexes such as issues, building space, and life cycle phase to identify automatically the chunks that are the most useful to the architect. CASECAD enables designers to retrieve previous design cases based on formal specifications of new design problems (Maher 1997).

Case-based reasoning can be divided into two phases: case retrieval and case adaptation. Case retrieval is more closely related to this research. Several techniques have been proposed for retrieving previous design cases. These include Bayesian networks (Dingsøyr 1998), conceptual graphs (Gerbé 1997), fitness functions (Altemeyer and Schürmann 1996), constraints (Bilgic and Fox 1996), object-based representation of cases (Maher and Gómez de Silva Garza 1996), and indexes of issues (Domeshek and Kolodner 1993).

Model-based reasoning tools use both general domain knowledge as well as knowledge from specific cases (for example Bhatta et al. 1994). These tools enable knowledge retrieval and reuse based on *a priori* set representations that are specific to narrowly defined domains and media types. IDEAL is a model-based reasoning tool that uses both general domain knowledge as well as knowledge from specific cases (Bhatta et al. 1994).

---

[3] Some research in case-based reasoning is more geared towards design assistance, relying on the human designer to guide the processes of case retrieval and case adaptation.

This research contrasts with the above efforts in that it is centered on the human designer and the natural reuse process as it is observed in professional practice. As a consequence, my approach is to support interaction with a corporate memory of less formal knowledge, rather than formal representation of cases and automatic case retrieval. Increasingly, research in case-based and model-based reasoning is converging with the approach adopted in this research that computer systems should support rather than automate design reuse (Simoff and Maher 1998, Popova et al. 2002).

Reuse Models
Several reuse models have been proposed, most of them in the field of mechanical engineering (Sivaloganathan and Shahin 1999 gives an overview). One model (Duffy et al. 1995) decomposes the reuse process into three processes and six knowledge resources (Figure 5).

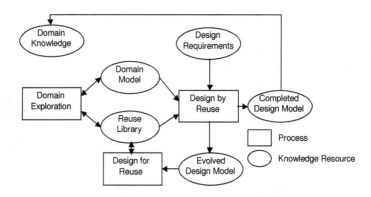

Figure 5: A design reuse model (Duffy et al. 1995).

The emphasis of the reuse model in Figure 5 is different from ours. Duffy et al. (1995) distinguish between design *by* reuse and design *for* reuse. They describe design for reuse as "the extraction of possible reusable knowledge fragments and the enhancement of their knowledge content". This is equivalent to the knowledge refinement step in the knowledge life cycle (Figure 1). In this research, there is a single knowledge resource: the corporate memory, which combines all six knowledge resources in Figure 5. This research focuses specifically on the designer's interaction with the corporate memory.

Other models are based around the phases of the product design process, from establishing specifications and requirements through to developing production plans (Shahin et al. 1997). For the purposes of this research, distinct phases of the design process do not matter as much as the evolution that a particular component goes through as it evolves from a conceptual idea to a fully specified physical entity.

Some reuse models recognize the negative effects of reuse. Design "overuse" (Lloyd et al. 1998) has been linked to the classic problem of design fixation.

Code Visualization and Reuse
In the field of software design, code reuse is an active research topic. A small subset of these efforts is dedicated to the development of applications that use visualization to assist in the retrieval and reuse of reusable software components. Table 2 gives some examples and compares them to this research.

Table 2: Related research in software reuse.

Project	Why?	What?	How?
This research	Find, understand → reuse	Projects, discipline subsystems, components (hierarchy)	Treemaps, fisheye views, node-link histories
MODIMOS	Monitor, reuse	Software components, class hierarchies (hierarchy)	Treemaps, node-link diagrams
Dali	Understand → reuse	Files, functions, variables (network)	Node-link diagrams
Vizbug++	Understand → debug	Program execution events (network)	Node-link diagrams
Jerding et al. 1997	Understand → reuse, reverse engineer	Interactions between classes, objects, functions, etc.	Time-series graphs, node-link diagrams, various others
CodeBroker	Find, understand → reuse	Software components	Latent semantic indexing (not visual), information delivery ("push" rather than "pull")

MODIMOS (Zieliński et al. 1995) allows the designer to monitor software applications made up of heterogeneous components, and indirectly supports reuse. It uses both node-link diagrams as well as treemaps for visualizing hierarchical structures such as class hierarchies.

Dali (Kazman and Carrière 1998) visualizes software systems using networks of files, functions, and variables (the nodes), as well as relationships between them (the links). They propose operations such as aggregation for reducing the complexity of these displays.

17

VizBug++ (Jerding and Stasko 1994), with an emphasis on development rather than reuse, also uses node-link visualizations of networks of events such as *class define* or *instance create*. Ware et al. (1993) extend these ideas from 2D to 3D.

Jerding et al. (1997) propose the use of animated node-link diagrams and time-series graphs to visualize interactions in program executions.

All these projects emphasize the importance of the *understanding* of archived components (Jerding and Stasko 1994, Kazman and Carrière 1998). Retkowsky (1998) lists the steps for software reuse as finding, understanding, adapting, and integrating.

CodeBroker (Ye and Fischer 2002) is a code reuse system that autonomously suggests code fragments for reuse as the designer works.

Other Points of Departure
It has been noted above that this research is based on ProMem and SME. This research also uses ideas from the fields of information visualization and information retrieval (Figure 6).

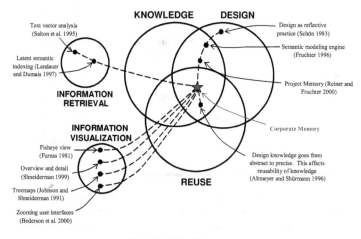

Figure 6: Points of departure.

Information Visualization
Information visualization has been defined as *the use of computer-supported, interactive, visual representations of data to amplify cognition* (Card et al. 1999).

18

This research will rely heavily on visualization techniques to support the finding of reusable knowledge and its presentation in context. The relationship between visualization and knowledge reuse is closely linked to the relationship between visualization and creativity, explored by researchers in human-computer interaction (Burleson and Selker 2002). Shneiderman (1999, 2002) identifies the ability to rapidly explore large amounts of information as an important step in creative processes. The main principle identified by Shneiderman is "overview first, zoom and filter, and then details-on-demand" (for short: *overview and detail*).

In this research, design knowledge, i.e. the knowledge captured by SME, is predominantly hierarchical in nature i.e. level of granularity trees and version trees. Several techniques have been developed for visualizing hierarchical information[4]. These techniques can be divided into two categories: those using *connection* (traditional node-link diagrams, e.g. Robertson et al. 1991) and those using *enclosure* (treemaps, Johnson and Schneiderman 1991). Treemaps are particularly effective for visualizing large hierarchies in a limited space.

Shneiderman's principle of *overview and detail* relates to *interaction*. A similar principle that relates to the *visualization* itself is *focus and context*. This principle states that the user simultaneously needs both an overview as well as detailed information, and that these can be combined in a single display (Card et al. 1999).

In this research, the principle of *overview and detail* will be used to support the designer in identifying a potentially reusable item from the corporate memory, and the principle of *focus and context* will be used to support the designer in exploring this item's context.

The fisheye view (Furnas 1981) is a *focus and context* visualization that allows the user to *zoom in semantically* on an item while keeping its context in view. This combination of local detail and global context would support the designer, not only in understanding the context of the item that he/she is considering reusing, but also discovering whether there is a related item which is also (or more appropriately) reusable.

Zooming user interfaces (see for example Perlin and Fox 1993) address the problems of limited screen space. They have been shown to be more effective than their non-zooming counterparts for many applications, including image browsing (Combs and Bederson 1999) and web browsing (Bederson et al. 1996).

Information Retrieval
To help guide the designer's exploration of the corporate memory, it will be necessary to quantify the similarity between projects, or the relevance of any item given the problem the designer is working on. An SME project memory can be thought of as a

---

[4] An extension to tree hierarchies is a multitree (Furnas and Zacks 1994).

structured (hierarchical) set of semantic keywords or annotations. Several techniques have been proposed for comparing texts in the field of information retrieval. One example is *text vector analysis* (e.g. Salton et al. 1995), in which a text is represented as a vector in high-dimensional space, with each dimension representing the frequency of a word in the text. Texts can be compared by calculating the distance or angle between their vectors.

Latent semantic analysis (Landauer and Dumais 1997) is a refinement of text vector analysis. The principle behind latent semantic analysis is that the way a pair of words occurs in small sub-samples of language reflects the "psychological similarity" between those two words. This similarity can be deduced by reducing the dimensionality of the text vector. Using this technique, a search for "cooling tower" would also return results with the term "piping". A less experienced designer might not know that a common problem in cooling towers is the routing of piping, but the system would infer this from the way that these terms repeatedly appear together in the corporate memory.

Chapter 3

ETHNOGRAPHIC FINDINGS: DESIGN KNOWLEDGE REUSE IN THE ARCHITECTURE, ENGINEERING AND CONSTRUCTION INDUSTRY[5]

Introduction

This chapter presents the results of an ethnographic study of practitioners in the architecture, engineering, and construction (AEC) industry. The objective of this study is to investigate the process of *knowledge reuse* by AEC practitioners as a precursor to designing a computer system that would support this reuse[6].

In this research, knowledge reuse is defined as the reuse of knowledge from previous completed (or "dormant") projects in a current (or "active") project. In particular, this study focuses on *design* knowledge reuse, i.e. the reuse of *designed artifacts or artifact subcomponents* from project to project. Our observations indicate that design knowledge reuse is one of the most common types of reuse, and an area with great potential for support by a computer system.

In many firms, one of the primary mechanisms for knowledge reuse is through mentoring relationships where a novice goes to an expert with questions. In general, this relationship is very effective and should not be threatened by a computer system. In this study, special attention was paid to these mentoring relationships in order to understand how a computer system can support (rather than replace) mentoring.

Scope of this Study

This study was conducted for the purpose of designing a computer system for design knowledge reuse. The aim of using ethnography for technology design should be to understand practitioners' needs in order to design technology to meet those needs, rather than starting with a technology and trying to understand how this technology can be used in a certain setting.

The idea of a reuse system is abstract enough to allow observations of the practitioners' needs and respond to those needs in the design of a system. In the course of this study, many other areas where technology could improve the working lives of AEC

---

[5] The contents of this chapter were first published in Demian, P. and Fruchter, R., 2006. "An Ethnographic Study of Design Knowledge Reuse in the Architecture, Engineering and Construction Industry" Journal of Research in Engineering Design, volume 16, number 4, pp. 184-195. They are reproduced here with permission of Springer.

[6] For a discussion of the use of ethnographic methods for design, see Blomberg et al. (1993). Lloyd et al. (1998) conducted a similar study of a small manufacturing and design organization, looking specifically at design overuse. Bucciarelli (1994) uses ethnographic methods to study collaborative design in three engineering design firms.

practitioners were encountered; however this research focuses specifically on design knowledge reuse.

Method

Data for this study was collected through interviews with and participatory workplace observations of AEC practitioners. The vast majority of the ethnographic data collected was centered on a structural design office of Z Inc (pseudonym) Structural Engineers and Builders in Northern California. The firm has three offices in the US with a total of twenty engineers. The California office employs five engineers, including the founder and senior engineer of the company. A two-week field study of this office was conducted in June 2000. Observations were recorded by taking notes throughout the working day. During this two-week period, three project design meetings were held, each lasting for about three hours. All three design meetings were video recorded in their entirety. During this period, Two engineers were accompanied on a site visit to a hotel construction site in Southern California. This site visit was video recorded. In the two years following the field study, several return visits were made to the design office to interview the engineers and make further observations. These meetings were audio recorded and transcribed.

In addition to the Z Inc study, four further interviews were conducted with AEC practitioners from other companies in April 2002. These interviews were audio recorded using a laptop computer and transcribed. Each interview lasted for approximately half an hour. Two of the interviews took place in the workplace of the informant, in those cases the informant offered to give a tour of his/her office or cubicle. Of the four informants, one was an architect, one was a structural engineer, and two were construction managers. Of those four informants, two were experts (with more than 15 years of experience), and two were novices (with less than five years of experience).

All the gathered data (notes from observations, transcripts, and documents) were analyzed qualitatively. Instances of design knowledge reuse were identified and coded. In particular, the analysis focused on two aspects:

- **Mentoring**. The senior engineer at Z Inc, an experienced designer with more than twenty-five years of experience, played a very important mentoring role. Special attention was paid to the interactions between this senior engineer and the novices who came to him with questions, and to the way in which he reused knowledge from his personal experiences when answering these questions.

- **Company standards and typical building details**. At the time of the study, Z Inc was in the process of developing a company standards system. The majority of these standards are typical building details, but the standards also include spreadsheets, document templates, and work protocols. Many of the discussions with the engineers at Z Inc were devoted to talking about the company standards.

Construction Managers

Design knowledge reuse does not appear to be a pertinent issue for construction managers. Constructions managers generally generate a lot of paperwork. These are usually workflow forms such as *requests for information* (RFIs). If a part of the design documents is unclear, the subcontractor responsible for this part of the building submits an RFI to the general contractor, who forwards it to the appropriate member of the design team.

Construction managers deal with huge volumes of these forms. The forms are usually kept in paper format, although computer systems are frequently used to help manage and track them. One of the construction managers interviewed revealed that the project she was working on had generated over 3500 RFIs so far. She showed us a huge binder full of them. It appears, however, that such forms are of little use after the project is over. When asked whether she would ever refer back to those records after the project was completed, this construction manager replied that she would only do so in the event of a problem arising in the completed building within the one-year guarantee period offered by her company. After this period, records from the project are usually sent to a huge warehouse in a nearby city.

Both construction managers interviewed agreed that only a few "standard" documents are reusable from project to project. A young construction manager noted that her company maintained a database of such documents, but when describing her day to day work earlier in the interview, she never mentioned using this database. An experienced construction manager gave us two specific examples of document templates that he frequently reuses from project to project: a Storm Water Pollution Prevention Plan, and a Traffic Control Plan.

This experienced construction manager said that, in his opinion, the form of knowledge reuse that would be the most useful to a construction practitioner is not document templates but *cost information*. A large part of the job of the construction manager is to estimate the cost of a construction project, often when the design is still at a very early stage. An experienced construction manager does not rely completely on published cost estimates, but keeps track of *actual* cost data from previous projects and uses that information to improve the accuracy of future cost estimates.

Finally, both construction managers interviewed acknowledged that experiences from previous projects played a large part in selecting subcontractors for current projects.

To summarize, even though construction managers are becoming involved increasingly early in the design process, they do not consider design knowledge reuse (i.e. the reuse of designs) to be an integral part of their professional practice. Perhaps the kind of knowledge that they *do* reuse can more accurately be described as *domain expertise*, which falls outside the scope of this study.

Designers: Architects and Engineers

In contrast to construction managers, the designers (architects and engineers) interviewed were more aware of reusing knowledge from past projects in their work. For both architects and structural engineers, knowledge reuse frequently takes the form of reusing standard building details[7]. All the designers interviewed emphasized the importance of understanding a detail before using it in a new project. They were quick to point out that designing a building involves much more than putting together standard building components.

Reuse by designers is not limited to designed building components. Designers, particularly structural engineers, frequently reuse spreadsheets and other design tools such as structural analysis models. The Z Inc structural engineering office included "standard spreadsheets" in its database of company standards. At another structural design office, the designer said that she had accumulated a small personal collection of spreadsheets during her nine months at the company. She also added that she frequently refers back to structural analysis models from previous projects to check the assumptions she made because she had to model a similar situation in her current project.

To summarize, designers generally reuse knowledge more frequently than construction managers. The remainder of this chapter looks more closely at this reuse: what are the mechanisms by which it occurs and what are the specific types of knowledge reused?

Knowledge Reuse Through Social Knowledge Networks

Two distinct attitudes to knowledge reuse were observed. The first attitude is that knowledge could (and *should*) be captured and stored in an external repository for all employees to share and reuse. The second and more common attitude is that the best sources of knowledge are the people in the company, who often possess a great deal of tacit and contextual knowledge that is difficult to encode and capture. Companies that adopted this attitude considered the role of technology to be to help cultivate and leverage *social knowledge networks*[8].

---

[7] A standard detail is a small part of a building design that changes very little from project to project, for example a detail for joining a beam to a column. Designers produce a set of drawings as the output of the design process. Several sheets of these drawing sets are taken up by typical or standard details.

[8] The term *social knowledge network* is used here informally. The network is *social* in the sense that it consists of people. The links between the people are each individual's set of contacts to whom that individual goes with questions. It is a *knowledge* network in the sense that the person on one end of the link is a knowledge seeker, and the person on the other end is a knowledge provider, and so knowledge flows through the network. Several other terms have been proposed for describing similar or related phenomena.

- Organizations can be viewed as consisting of individuals interconnected as members of *social networks* (Zack 2000).
- *Communities of practice* are groups of people with similar goals and interests, exposed to a common class of problems (see for example Wenger 1998).
- The process of *transactive memory* was originally studied in personal relationships (Wegner 1987) and later extended to people in work situations (Hollingshead 1998). The basic idea is that a group of people working together forms a shared understanding of each individual's knowledge. New information is directed to the person whose expertise will facilitate its storage. When knowledge is needed, it is retrieved based on the relative expertise of the individuals in the transactive memory system.

24

These social knowledge networks are naturally fostered through social events and protocols at companies. When asked how she learnt the necessary skills for her job, a young construction manager described a training program offered by her company:

They do have a training program at Albertson Construction. Every new employee has to go through it. It gives you just enough information to get started. I learned some things in the program, but the really important thing I got out of it was the business cards of the people who were teaching the program, whom I could call with questions.

The training program helped, but the real benefit was the knowledge network: knowing who to ask and who knows what in the company. Similarly, a young engineer highlighted the importance of social knowledge networks in her company, and the conscious efforts of the management to promote these networks[9]. She described her office as a supportive environment where colleagues were always willing to help:

There are so many people in this office, and they are all really nice and approachable, I just know who to ask. You know, this guy is really into nonlinear analysis... We have a lot of lunchtime meetings about miscellaneous subjects, and you just hear whoever speaks up... you can tell who is into what...

Even when the information is available in some external repository, the practitioners interviewed indicated that they often rely on the social network to help them locate information in this repository.

In some cases, software systems were encountered that were intended specifically to support social networks. At one company, each employee is invited to submit an online profile listing his/her skills. People at the company are encouraged to search these profiles and locate useful contacts whenever they have a question. At an architectural firm, an online database of project profiles is maintained. Each profile contains a brief description of the project: the type of building, the budget, the location, and *the people who worked on the project.* When speaking to an architect from that company, he reported that the most useful aspect of the project profiles system is the ability to locate people in the company that have worked on similar projects.

The idea is that these [project profiles] would be sitting on a web site, an intranet, and would be available for teams, so that they could say, "Gee, who has done this type of building before." You could go up and find, oh,

---

- The importance in the workplace of *personal social networks* that cross traditional organizational boundaries has been recognized, and so has the effort required to create and maintain such networks (Nardi et al. 2000).

[9] Interestingly, this company prides itself on its ability to retain employees in the company, and to support its employees' learning and training aspirations.

that was studio X, and you find out which people in studio X, and you could call them.

To conclude, social knowledge networks are a crucial mechanism by which knowledge reuse occurs in current AEC practice. AEC practitioners prefer to ask colleagues who have worked on similar projects or have been faced with similar problems. Even when the information being reused is externally encoded (e.g. an old blueprint), the social knowledge network is relied upon to help identify, locate, retrieve, and understand this information.

Internal Knowledge Reuse: The Importance of Context
It is useful to distinguish between *internal* and *external* knowledge reuse:
1. *Internal knowledge reuse*: a designer reusing knowledge from his/her own personal experiences (internal memory). For example, a structural designer might remember that the last time she designed a floor slab for a hotel ballroom it was too thin, which resulted in vibration problems. The next time she is faced with a similar design situation, she designs the floor slab to be deeper.

2. *External knowledge reuse*: a designer reusing knowledge from an external knowledge repository (external memory). For example, the same structural designer might look for floor slab designs in her company's standard components database. She retrieves a floor slab design that comes with a spreadsheet for calculating the correct slab thickness. This spreadsheet takes into account the company's previous experiences with vibrating floor slabs and increases the depth beyond the minimum required by the building code.

The effectiveness of reuse through social knowledge networks can be partly attributed to the fact that it relies on internal (rather than external) knowledge reuse. When answering questions from colleagues in the knowledge network, the experienced AEC practitioner invariably refers back to his/her own experiences. During this study, many observations were made of the interactions between the senior structural engineer at Z Inc (an expert structural designer) and novice designers at the office in order to understand the process of internal knowledge reuse, i.e. how the expert "interacts" with his own internal memory when answering the novices' questions.

The senior engineer's internal knowledge reuse process was observed to be very effective[10]. He was always able to recall directly related past experiences and apply them to the situation at hand. Two key observations in particular characterize the effectiveness of internal knowledge reuse:
1. Even though the senior engineer's internal memory was very large (he has over twenty-five years of experience), he was always able to *find* relevant designs or experiences to reuse.

---

[10] It would have been impossible to evaluate his mental retrieval process quantitatively in terms of precision and recall.

2. For each specific design or part of a design he was reusing, he was able to retrieve a lot of *contextual knowledge*. This helped him to *understand* this design and apply it to the situation at hand. When describing contextual knowledge to the novice, the senior engineer explored two contextual dimensions: the *project context* and the *evolution history*.

The *project context* dimension encapsulates the levels of granularity at which contextual knowledge about the design project can be explored. Given an item from a past project, the following directions of exploration were identified:

- *UP: From component to subassembly.* Designers move upwards along this dimension to explore the discipline (or building subsystem) and project in which this item occurs. This is best explained using an actual scenario that was observed at Z Inc when a novice designer asked the senior engineer how to go about designing a cooling tower frame[11]. The senior engineer identified a cooling tower frame from a previous project that the novice could reuse. He explored the project context upwards by recalling the structural system and even the entire project from which this cooling tower frame was taken.

- *DOWN: From subassembly to component.* Designers move downwards along this dimension to consider the subparts or subcomponents of which this item is composed. The senior engineer explored the project context downwards by describing some of the interesting beams, columns, braces, and connections of which the frame was composed.

- *SIDEWAYS: From one item to related items.* Designers move sideways to explore related items in the same project or from other projects. The senior engineer explored the project context sideways by considering the cooling tower unit (a related item) supported by the frame to determine what load it exerted on the frame.

The *evolution history* is the record of how an item evolved from an abstract idea or a set of requirements to a fully designed physical entity. Given an item from a previous project, the following directions of exploration were identified:

- *UP: From detailed to conceptual.* Designers move upwards along this dimension to trace the concepts that were explored early on in the design of this item. The senior engineer explored the evolution history upwards by showing the novice a sketch of the conceptual braced frame design that was created early in the design process.

- *DOWN: From conceptual to detailed.* Designers move downwards along this dimension to follow the evolution of this item into a fully designed physical component. The senior engineer reusing the frame described its evolution into a

---

[11] In this context, a cooling tower is a large air conditioning unit. A cooling tower frame is a support structure that holds the cooling tower up.

27

fully detailed design in a CAD file, and even showed the novice photographs of the frame as built.

- *SIDEWAYS: From alternative to alternative.* Designers also move sideways to explore the different alternatives that were considered at any stage in the design process. The senior engineer reusing the cooling tower frame recalled that steel and concrete alternatives were considered. He told the novice that perhaps the concrete alternative that was originally abandoned could now be reused.

From the cooling tower scenario described above and many others like it that were observed, the following formalizations of the process of internal knowledge reuse can be made:

- The process of internal knowledge reuse can be summarized into three steps:

   1. *Finding* a reusable item.

   2. Exploring its project context in order to *understand* it and assess its reusability.

   3. Exploring its evolution history in order to *understand* it and assess its reusability.

- There are therefore six degrees of exploration, three – up, down and sideways – in each of the two contextual dimensions (project context and evolution history).

These observations of internal knowledge reuse can be used as the basis for supporting external knowledge reuse from an external knowledge repository.

External Knowledge Reuse: Company Standards and Typical Details
At the Z Inc office where the field study was conducted, mentoring relationships play a large part in promoting knowledge reuse, where the experienced senior engineer uses his own process of internal knowledge reuse to guide and instruct the less experienced designers at the company. However, at the time of the study, Z Inc was investing a large amount of resources into developing a software system to support reuse. In contrast to the reuse systems observed at other companies, this system was designed to support directly external knowledge reuse, rather than reuse through social knowledge networks. Specifically, Z Inc was developing a web-based system for company standards.

This system is in use at the time of this research. The majority of the standards in the system are typical building details, but the standards also include spreadsheets, document templates, and work protocols. Figure 7 shows a screenshot of the standards web interface.

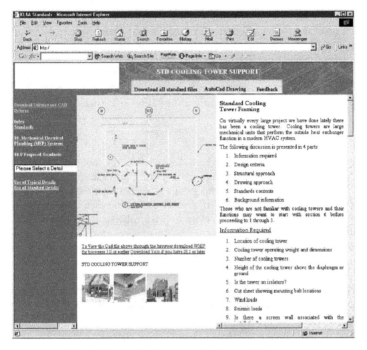

Figure 7: A screenshot from an engineering standards system.

The young engineers intervuewed reported being generally satisfied with their experiences using the standards database, although the success of the system depends largely on the interactions of the young engineers with the senior engineer. The major problem with the standards system is that it does not adequately support the two activities observed during internal knowledge reuse: *finding* reusable items and *understanding* these items in context.

The standards are arranged in a two level hierarchy. The top level categorizes the standards by material (steel construction, concrete construction, wood construction, etc.) and the next level by the type of standard (sample drawings, typical details, design guides and tools, etc.). The senior engineer, who helped design the system, acknowledged that it is difficult for the novices in the office to find useful standards:

> It's pretty much a search assisted by the broad categories... you know, they are divided by material... And so if you're designing a steel or a concrete building, then you go look in the concrete section. It's rather interesting that I can get in and out pretty easily, because I have in my head a relatively refined search algorithm already. It turns out that the kids [i.e. the novices] don't.

29

The young designers interviewed to all agreed that it is easy to find a specific standard in the system if they know *a priori* which standard to reuse, for example if they have reused this standard in the past and are aware of its existence. The real problem is in the situation where they do not know what they were looking for, only that it should be a standard that is relevant to their current design task[12]. In those cases, they often rely on more senior designers to help them identify and find a useful standard:

> *Sometimes I'll ask Eric or Frank if they know of something that's been previously done, if they know were it's at, because I've probably not experienced it in the short time I've been here.*

The same problem applies to standard designs that are not necessarily from the standards system, but that have been created during previous projects. When asked whether he would reuse designs from previous projects only if he had worked on these projects, a young designer replied, "If it's not something that I have done myself, I won't know to look for it." Again, interaction with the senior engineer is an integral part of reusing from previous projects:

> *If I am generating a detail for the rolling door... Eric [the senior engineer] would ask me, "What are you working on?" I would say the rolling door. He'd say, "Well, we had one of those on this job." And we've done that before, we've actually gone back two or three jobs back, jobs I haven't worked on, and looked for a detail, and found it, or said, this is similar, let me use it and modify it. So the company memory goes back further than me, but it goes back as far as Eric.*

Young designers are usually unable to find a reusable item from the standards system or previous project archives without having been previously exposed to this item, or interacting with the mentor who guides them on what to reuse. The second reason for the ineffectiveness of external reuse systems is that the design knowledge they offer is *decontextuallized*. One of the young designers at Z Inc reported that he often had to ask the senior engineer questions about a standard because "he [the senior engineer] did a lot of them... he's dealt with a lot of them personally". There is a lot of contextual information missing from the standards, contextual information that can only be provided by the senior engineer, who has helped to develop the standard and who has probably worked on the project for which that standard was originally designed. Again, this contextual information falls along the *project context* and *evolution history* dimensions.

The importance of the *project context* when reusing a detail becomes very apparent when the senior engineer discusses the tradeoff they had to make in the design of each

---

[12] Ye and Fischer (2002) make the similar observation that users are often unable to utilize reuse systems because they are unaware that there is something relevant in the system, or they don't know what to look for and so are unable to formulate a query. Their solution involved *implicit queries* combined with *information delivery* (information is pushed by the system rather than pulled by the user).

standard between *knowledge-rich* standards that were very *specific* and *generic* standards that were nevertheless applicable to a wide range of projects. Once a typical detail is taken out of its project context and standardized, it loses most of its value. One example offered by the senior engineer is the disagreement he had with the editor of the standards over the standard for an elevator pit:

> *Another one was an elevator pit. The one that we put on the standards was one from the LA project, which really wasn't a standard, it was totally special. Bart [a retired engineer with about fifty years of experience who was put in charge of editing the standards] rightfully said it was too special. So he threw that one out and he proposed one which was his detail. It was really innocuous, stripped of any specialized information at all. Bart is very much old school in that a building is just an assembly of details, and that there's nothing wrong with drawing one detail and completely ignoring the fact that there is another detail that must interface with it. He just draws all of these details independently and expects the contractor to figure out how they all fit together. Now in an elevator pit you have…[goes on to describe the components of an elevator pit]… So you have all these things happening in an elevator pit. Bart's detail shows a floor and a wall, because that's the simplest form of an elevator pit, as far as he's concerned. And furthermore, if you show a wall and a floor, it's symmetrical and there's no point drawing the other half, so he only shows half of it! One wall and half a floor! Which makes his detail look exactly like our slab stair detail! [i.e. not what it is supposed to look like]. And I find it offensive because it doesn't look like a pit. Our detail shows both walls because one of them is the back wall of the elevator, and has a solid wall, and the other one is the front wall, and that has a sill detail. Bart doesn't want to show the sill, because that's a different detail, somewhere else in the drawings…Bart still likes to do all of his details as disembodied little pieces. If you put his details together you don't actually have a whole because there's a whole bunch of knowledge that goes in there that he expects someone else to fill in.*

Just as important as the project context is the *evolution history* behind a design. When asked what information is missing from the standards, a young designer replied that he needed to know the function and rationale driving the development[13] of the design:

> *Usually, it's the purpose behind the design, or the reason behind developing the design the way it was… what was the person thinking when they developed the standard, that's the key thing.*

---

[13] I am referring here to the process of developing a design from an abstract idea or requirement to a precisely specified physical component. There is a macro evolution process that occurs when a design is reused from project to project and is improved and refined each time it is reused. This idea will be addressed later in this chapter.

31

Invariably, when instructing the young designers to reuse a component from a previous project, the senior engineer would mention some relevant facts about the evolution of this component when it was originally designed. This information usually had important implications for whether or not (and how) this component would need to be modified before it could be reused. For example, when instructed to reuse a frame from a previous project that was located in Las Vegas in a current project that was located in Illinois, a young designer rightly noted that the members of the frame would probably be too small because Illinois is a high-wind area whereas Las Vegas is not. The senior engineer replied that the original design was "conservative for Las Vegas, so it would be ok for Illinois."

To summarize, external reuse systems fail because they do not support the activities that were observed to make internal reuse effective: the ability to *find* and *understand* reusable items. The partial success of the standards system at Z Inc can be attributed to the important role played by the senior engineer. He initiates most of the design reuse by directing the young designers to useful standards in the system. Having been personally involved in the development of these standards, he is able to provide a lot of contextual information that ensures that these standards are effectively reused; this is contextual information that is not directly available from the system.

Reusing Designs: Productivity versus Creativity
Reusing items from previous designs can increase the productivity of the design process, but may also compromise the creativity of the designed artifact. This tradeoff between productivity and creativity was observed both when talking to the designers at Z Inc about the standards system and when observing their reuse activities during design meetings.

Two general characteristics of a design item are considered by the designer when making a reuse decision: *level of granularity* and *level of abstraction* or *precision*.

The *level of granularity* is the size of the design chunk being reused, from the whole artifact to small subcomponents of the artifact[14]. Reusing small "chunks" of designs, while not very helpful in increasing the designer's productivity, is less likely to compromise the creativity of the artifact being designed. In the AEC industry, the reuse of *standard details* from one project to a completely different project is not uncommon. A standard detail is, by definition, a small chunk and can be used in a wide variety of design situations without compromising the creativity of the new design. In fact, at Z Inc, as in other design practices observed, importing details from other projects is a standardized task in the design process (or perhaps more accurately: the process of preparing drawing sets). On the other hand, large chunks of design, while inherently richer in knowledge, are less reusable. This is evident from the episode cited above

---

[14] Fruchter (1996) recognizes level of granularity as an important factor in capturing and reusing design knowledge.

32

where the senior engineer proposed a standard for an elevator pit that was totally specific to the project for which it was originally designed. The editor of the standards, as part of his process of making this standard more generic and applicable across a variety of projects, trimmed down the level of granularity of the standard to focus on the essential subcomponents of an elevator pit, resulting in what the senior engineer termed "a disembodied little piece" of a standard.

The important point to make is that an experienced designer will manage the tradeoff between productivity and creativity by reusing as large as possible a chunk of design, given the differences between his/her current design task and the original situation for which the design being reused was generated. Returning to the cooling tower frame example, when the senior engineer instructed the young designer to reuse the cooling tower frame, the young designer objected that it would be inappropriate to reuse the entire frame because it was part steel and part concrete. The senior engineer replied that it was still possible to reuse just the steel part in the current project (i.e. reusing a smaller chunk).

The *level of abstraction* or *precision* is the degree to which a design has evolved from an abstract or conceptual idea to a precisely defined physical component[15]. For example, a structural frame design will usually evolve from an abstract concept (a sketch of an eccentrically braced frame) through a developed design (a CAD drawing with approximate dimensions and all members represented as centerlines) to a detailed design (a 3D CAD drawing with actual member sizes and connections between members). Designs closer to the abstract end of the spectrum are more generally reusable, but bring about only a small increase in productivity because the design still needs to be developed. However, if the reuse is occurring early on in the current design process, then this does not pose a problem.

For example, the reuse of the cooling tower frame mentioned above occurred relatively early in the design process of a hotel project. When the young designer raised objections about reusing the cooling tower because it came from a completely different type of hotel project, the senior engineer instructed her to "put something there as a placeholder, the dimensions and member sizes don't really matter right now". In other words, the senior engineer intended for the young designer to reuse the cooling tower at a slightly higher level of abstraction than that of a precisely defined cooling tower.

These two dimensions, level of granularity and level of precision, when used to define a two-dimensional knowledge space (Figure 8[16]), can be used to express the tradeoff

---

[15] Altmeyer and Schürmann (1996) refer to this as *refinement level*, and present a formalization of the design process in which each design step takes the artifact from a more abstract refinement level to a more precise refinement level. Similarly, Sutherland (1963) cited in Luth (1991) describes the design process as "a spiral that proceeds from the abstract to the particular over time."

[16] A similar diagram as that shown in Figure 8 is used by Rasmussen (1990) to represent the problem space in computer troubleshooting.

between productivity and creativity. At the top right corner of the knowledge space, the knowledge being reused is precise and pertains to a whole artifact, e.g. reusing a fully designed structural system for a building. This occurs, for example, during evolutionary design, which was not observed to occur frequently during this study, but may occur in other design domains. In this situation, "a lot of reuse is happening", but this is more likely to result in a loss of creativity since the whole artifact is used "as is", without exploration of alternatives.

Conversely, at the bottom left corner of the knowledge space, the knowledge being reused is abstract, and pertains to small subcomponents of the artifact (e.g. reusing an abstract principle for joining a beam to a column). In this situation, "less concrete reuse is happening", and will not affect the originality of the solution.

This tradeoff between abstract/finely-grained/reusable and precise/large-grained/unreusable was encountered by the Z Inc engineers during the design of the standards system. They quickly realized that for many components, there is no "standard way" to design that component, and that the actual design would depend on the context or the design situation. Their solution is to keep the standards as abstract as possible and as finely-grained as possible. As noted above, this makes the standard designs applicable across a wide variety of situations, but also strips them of much valuable contextual knowledge. When asked whether this made the standards system futile, the senior engineer replied, "No. The jobs themselves will motivate changes to the standard. We want to start from pretty much the same place before we start to diverge for every job... and every job *will* diverge."

This observed tradeoff has an important design implication: exploring the project context and exploring the evolution history of a design item being reused not only facilitate the understanding of this item and its effective reuse, but also help to manage the tradeoff between productivity and creativity.

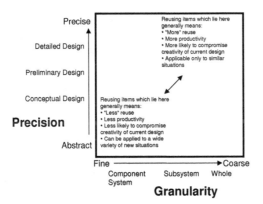

Precise

Reusing items which lie here
generally means:
• "More" reuse
• More productivity
• More likely to compromise
  creativity of current design
• Applicable only to similar
  situations

Detailed Design

Preliminary Design

Conceptual Design

Reusing items which lie here
generally means:
• "Less" reuse
• Less productivity
• Less likely to compromise
  creativity of current design
• Can be applied to a wide
  variety of new situations

Precision

Abstract

Fine ——————————→Coarse

Component   Subsystem   Whole
System

Granularity

Figure 8: The tradeoff between productivity and creativity.

The Role of the Mentor

What is the role of the mentor in relation to an external knowledge reuse system? It has already been noted that the senior engineer at Z Inc plays an important role in the effective use of the standards system by the young designers at the office. He is frequently the one who identifies the standard that can be reused and is instrumental in providing contextual information about that standard[17].

Another striking role played by the senior engineer at Z Inc is that of an "editor" of the standards: he helps to decide what should be included and what should not, and he ensures that all the standards in the system are of a high quality. When asked how frequently he himself uses the standards, the senior engineer replied, "I use them frequently, but just as frequently I edit them. I can't use them without changing them." For him, the standards are not static but are constantly being refined and improved. "It's a complex dynamic system in that the state of the knowledge changes every time you use the knowledge." These comments highlight the importance of the standards system (and indeed any reuse system) as a *knowledge refinery*. Each time a standard is reused in a new project, it is improved and refined and becomes more valuable[18].

---

[17] Mentoring can be thought of as a special case of reuse through social knowledge networks. A useful analogy is the distinction between *authorities* and *hubs* in hyperlinked environments such as the web (Kleinberg 1999). An authority is a page linked to by many other pages (cf. a person in the social knowledge network to whom people frequently go with questions about a certain topic, or a mentor) and a hub is a page that links to many other pages (cf. a person in the social knowledge network who always knows whom to ask). However, my observations at Z Inc indicate that the mentor is much more than an authoritative node in the social knowledge network, but is important for proactively promoting knowledge reuse.

[18] The idea of *knowledge refinement* has already been proposed in the knowledge management literature. It is usually used to refer to the process of cleansing, indexing or standardizing that must occur before captured knowledge is added to a repository (see for example Zack 1999). This is a slightly different sense than the one intended here, which is refinement *through reuse*. The idea of refinement through reuse has been identified as an important item in the knowledge management research agenda (Venzin et al. 1998) but has otherwise received little attention.

35

Not only does the senior engineer himself refine the standards, he expects the young designers to think critically about a standard before reusing it and, as a result of this critical assessment, to propose refinements to the standard. To him, this is an important mentoring mechanism by which the young designers learn.

> *Whenever my guys [the young designers] use the standards... we almost always talk about what a better way is to do the standard. We talk about what the purpose of the detail is, and how it is accomplishing it... They are much more willing [than an experienced engineer would be] to just use what they have, without really critiquing it too much... If they think about that, all the things that the detail is supposed to accomplish, and they critically evaluate the detail, then they will learn a lot. That's how you learn.*

From the senior engineer's perspective, the standards serve as a mentoring tool by encouraging the young designers to improve and refine the standard designs. Whereas the novice is usually concerned just with the *outcome* of reuse (higher productivity), the mentor is also interested in the *process* of reuse, and its pedagogical value.

The senior engineer at Z Inc provides input on the extent to which a standard is applicable to the current design situation, and what modifications need to be made before the standard can be reused. Outside of Z Inc, almost all of the experienced practitioners emphasized that this is an area where young designers need a lot of support: knowing when and how an old design can be applied to a new situation.

The senior engineer at Z Inc encourages the young designers to use their own initiative to utilize the standards. The young designers in the office reported that their mentor usually expects them to have checked the standards database before coming to him with questions. Although they never hesitate to ask questions ("there's no such thing as a bad question"), they do think it is important to "do their homework" before taking the question to the mentor.

To summarize, a mentor can play the following roles in relation to an external design knowledge reuse system:
- An "editor" who decides what will be included in the system, and maintains the quality of the designs therein.
- A "coach" who encourages the young designers to think critically about the designs in the system and to learn from them and improve them.
- An "expert" who has first hand experience related to the designs in the system and can provide contextual information and input on what to reuse and how to reuse it.

Design Implications: Collaborative Versus Distributive

This study was conducted as a precursor to designing a computer system for supporting design knowledge reuse. We now turn to implications that the observations presented here will have on the design of this reuse system.

One of the issues that need to be resolved is the extent to which the reuse system should support reuse through social knowledge networks versus by retrieval from an external knowledge repository[19]. These two roles have been described as *collaborative* (bringing people together to facilitate knowledge flows between them) versus *distributive* (capturing knowledge in a repository and distributing it to users). The results from this study indicate that AEC practitioners frequently go to colleagues with questions, a result that is consistent with other published findings. For example, Allen (1977) studied engineers' information seeking behavior and found that their major source of information was direct communication with colleagues. However, this approach is not without its problems, for example regarding the accessibility of colleagues or the status implications of admitting ignorance (Gerstberger and Allen 1968). It can also be argued that this observed preference for asking people is a symptom of the shortcomings of archiving and reuse systems used in current practice.

When talking specifically about design knowledge reuse, it is clear that an external repository will be necessary because that is how designs are stored: in electronic CAD files or paper drawings, in electronic analysis models, in paper sketches and calculations, and so on. These types of content cannot be stored in, or retrieved from, a human colleague. Still, human mentors and colleagues do provide a lot of tacit and contextual information that is difficult to encode and store in a repository ("go and get the blueprint from the archive and I will explain it to you"). Perhaps the best approach would be to leverage both human knowledge and external knowledge repositories as far as possible: to capture and offer for reuse as much contextual information as possible, and at the same time to maintain a pointer to the human designers responsible for the design so that they can be contacted and asked for additional information not encoded in the repository[20].

This research focuses on the distributive aspects of knowledge reuse.

---

[19] Ackerman (1994) notes that information technology can support organizational learning in two ways: either by recording knowledge and making it retrievable, or by making individuals with knowledge accessible. Knowledge management systems that aim to make individuals with knowledge accessible take varying approaches. Some systems allow each individual proactively to create and maintain his/her social network. For example, ContactMap (Nardi et al. 2002) allows users to arrange their social networks in a visual map of individual contacts and groups. Other systems automatically mine sources of expertise information (such as e-mail archives) to infer the expertise and skills of individuals in an organization. One example is KnowledgeMail® by Tacit Knowledge Systems (http://www.tacit.com).

[20] See Ackerman 1994 for an example of this approach.

37

Design Implications: Finding and Understanding Knowledge in Context

The success of reuse through social knowledge networks and mentoring can be attributed to the fact that these rely on internal knowledge reuse, i.e. a human designer reusing knowledge from his/her personal memory or past experiences. Internal knowledge reuse, for those expert designers who have a sufficiently deep well of design experiences from which to draw, was observed to be very effective. It is effective because the designer can *find* reusable items and can remember the context of each item, which enables him/her to *understand* that item and reuse it effectively. An external knowledge reuse system should support these activities: *finding* and *understanding*. Understanding during internal knowledge reuse arises from recalling the project context and the evolution history of the item being reused. The external knowledge reuse system should therefore support project context exploration and evolution history exploration. This exploration would also help the designer manage the tradeoff between productivity and creativity.

If a reuse system will play a distributive role (so that reuse occurs by interacting with an external knowledge repository rather than with other humans) then it must also contain (insofar as this is possible) the kind of contextual information contained in a designer's internal memory. This notion is termed *knowledge in context*[21]. Knowledge in context is design knowledge as it occurs in a designer's personal memory: rich, detailed, and contextual. This context includes design evolution (from sketches and back-of-the-envelope calculations to detailed 3D CAD, analysis, and simulations), design rationale, domain expertise, and relationships between different perspectives within cross-disciplinary design teams. A *corporate memory* is a repository of knowledge in context; in other words, it is an external knowledge repository containing the corporation's past projects that attempts to emulate the characteristics of an internal memory, i.e. rich, detailed, and contextual. The corporate memory grows as the design firm works on more projects.

If the corporate memory is to contain knowledge in context, then the design knowledge should be organized by project so that the designer can understand the design being reused in the context of its original project and design process. This is in stark contrast to the standards system used at Z Inc, which contained decontextualized fragments of designs organized into abstract categories.

Design Implications: Support for Novices and Mentoring

The successful use of the standards system at Z Inc by the young designers was largely dependent on their interactions with the senior engineer. A reuse system should support

---

[21] This term has been used by Finger (1998) in a similar sense. She notes that designers must seek out previous designs in the context of a design problem. Design is a process of constructing a theory of the artifact, not merely constructing a manufacturable description. This *artifact theory* is a contextual theory that provides knowledge for describing and analyzing an artifact and for explaining and predicting the nature of the artifact.

reuse by novices in the absence of their mentor, but must also be able to support the mentoring relationship.

When it comes to design reuse, a novice with little design experience does not know what to look for and where to find it. A reuse system must be able to take some representation of the designer's current design task and generate some measure of relevance between the current design task and each item in the repository. This *implicit query* (Ye and Fischer 2002) can be extremely helpful to the novice whose unfamiliarity with the contents of the repository prevents him/her from formulating a useful query. On the other hand, a designer with more design experience, or who is looking for a specific item, perhaps one that he/she has worked on, should be able to formulate a query explicitly. These explicit queries can also be used by the mentor, or by a novice following instructions from a mentor, as part of the mentor's coaching activities.

A major role that can be played by the mentor is that of an editor of the contents of the repository. A reuse system must be able to act as a dynamic *knowledge refinery* that enables the designs contained therein to evolve and improve. This idea falls outside the scope of this research but is identified as an important direction for future research.

Closing Remarks
Knowledge reuse in current AEC design practice occurs largely through social knowledge networks. Even when reuse from an external repository occurs, a human expert is usually needed to provide proactive input on what to reuse and contextual information on the designs being reused. Both of these observations are attributed to the effectiveness of internal knowledge reuse, the reuse of knowledge from one's personal experiences. Internal knowledge reuse is effective because the designer can *find* items to reuse, and can recall the context of these items and can therefore *understand* them.

My reuse system will be a *corporate memory*, a rich, detailed repository of *knowledge in context*. The corporate memory will support *finding* and *understanding*. Understanding can be brought about by enabling the designer to explore the project context and evolution history of the found item. These explorations will also help the designer to manage the tradeoff between productivity and creativity, by facilitating reuse at the appropriate levels of granularity and abstraction.

The corporate memory must also act as a dynamic knowledge refinery rather than a static knowledge repository. Finally, it is important to acknowledge that knowledge reuse cannot occur solely by interacting with the corporate memory, but will probably happen in a social context, whether the designer interacts with colleagues or with his/her mentor. These points will be considered outside the scope of this research.

Chapter 4

DESIGN AND RESEARCH METHODOLOGY

In Chapter 3, the results from an ethnographic study of design knowledge reuse by AEC designers were described. This chapter describes how these findings were used to design *CoMem* (Corporate Memory), a prototype system for supporting design knowledge reuse in the AEC industry, and how CoMem was evaluated.

Summary of Ethnographic Findings and their Design Implications
Knowledge reuse in current AEC design practice occurs largely through social knowledge networks. Even when reuse from an external repository occurs, a human expert is usually needed to provide proactive input on what to reuse and contextual information on the designs being reused. These observations are attributed to the effectiveness of *internal knowledge reuse*, the reuse of knowledge from one's personal experiences. Internal knowledge reuse is effective because the designer can *find* items to reuse, and can recall the context of these items and can therefore *understand* them and reuse them appropriately.

This suggests that an external repository of design knowledge should, insofar as this is possible, emulate the characteristics of design knowledge as it occurs in the designer's internal memory[22]. In other words, the reuse system will be a *corporate memory*, a rich, detailed repository of *knowledge in context*. The system should support the same activities observed during internal knowledge reuse, i.e. the corporate memory should support *finding* and *understanding*.

To support finding, particularly for novice designers unacquainted with the contents of the corporate memory, the system should be able to generate some measure of relevance between the designer's current task and each item in the corporate memory.

Understanding can be brought about by enabling the designer to explore the *project context* and *evolution history* of the found item. These explorations will also help the designer to manage the tradeoff between productivity and creativity by facilitating reuse at the appropriate levels of granularity and abstraction.

Two important ethnographic observations will be considered outside the scope of this research and will not be directly addressed:
- *Knowledge refinement* through reuse is extremely important. The corporate memory must act as a dynamic knowledge refinery rather than a static knowledge repository.

---

[22] A discussion of the possibility, desirability, and consequences of capturing context digitally, is presented by Grudin (2001).

- Knowledge reuse cannot occur solely by interacting with the corporate memory, but will happen in a social context, whether the designer interacts with colleagues or with his/her mentor. The corporate memory must play a *collaborative* (as well as a *distributive*) role.

Tasks in Current Practice: Retrieval and Exploration

Two main kinds of reuse tasks were observed in current practice. *Retrieval* occurs when the designer is looking for a specific item: "I am looking for the cooling tower frame (component) from the structure (discipline subsystem) of the Bay Saint Louis Hotel (project) that we worked on five years ago". *Exploration* occurs when the designer has no idea what to look for, only that it should be a relevant item or that it should satisfy certain conditions: "I am stuck trying to design a hotel cooling tower, is there anything in the system that can help me get started?" In between the two extremes of retrieval and exploration there lie a whole range of tasks, for example when the designer might have some notion that there is a specific item in the system that would be helpful, but cannot remember exactly where it is: "I remember designing a hotel cooling tower a few years ago... what project was that for and where in the system can I find it?"

The dual reuse modes of retrieval and exploration apply to both *finding* as well as *understanding*. As previously noted, understanding can be supported by providing contextual information about the item being reused. If the designer is reusing an item with which he/she is completely unfamiliar, then he/she will probably explore the project context and evolution history. On the other hand, if the designer is somewhat familiar with the design being retrieved, then he/she might need to find a specific item of contextual information: "why did we decide to go with the braced frame instead of the moment resisting frame?"

Stakeholders in Current Practice: Novices, Experts, and Mentors

Three groups of stakeholders in an external reuse system can be identified from the ethnographic study: novices, experts, and mentors.

The novice is a young designer with less than five years of experience. He/she will have worked at the company for only a few years and so will be unfamiliar with the contents of the corporate memory. The novice is more likely to *explore* the corporate memory than retrieve specific items. The novice rarely knows exactly what to look for and so is unable to formulate an explicit query. Some measure of relevance between the novice's current task and each item in the corporate memory would be extremely useful in guiding the novice's exploration. Context is extremely useful to the novice, as he/she will probably use the corporate memory as a learning resource, and so the rationale or decision process behind a reusable design would be just as important as the design itself.

41

The expert is a designer with five to fifteen years of experience. He/she will be quite familiar with the contents of the corporate memory. The expert will consider the corporate memory to be a productivity tool rather than a learning resource. The expert is more likely to retrieve specific items from the corporate memory, although in a large company, the expert might find it useful to explore projects in which he/she was not involved. The expert might prefer to formulate explicit queries to search the corporate memory, and so he/she will rely less heavily on the relevance measure. Like the novice, the expert will need contextual information, even when retrieving specific items that he/she identified as reusable from memory. However, in this case, the expert will probably retrieve specific items of context, rather than explore the context in general. In terms of the entire knowledge life cycle, the expert will probably be a net producer (rather than consumer) of knowledge, and so from the expert's point of view it will be important to minimize the overhead for knowledge capture.

Finally the mentor is an expert designer with many years of experience. The mentor is responsible for managing and overseeing the design work of several expert and novice designers. In relation to the corporate memory, the mentor will be concerned about the quality of the designs in the corporate memory. He/she will want poor designs to be excluded or somehow marked as "poor". He/she will also want refinements to the designs to be captured in the corporate memory (i.e. knowledge refinement). The mentor can act as a "coach" who encourages the young designers to think critically about the designs in the system and to learn from them and improve them. He/she might occasionally direct novices to specific items in the corporate memory, but he/she would expect the novices to be able to interact directly with the system without his/her intervention.

Scenario-Based Design Methodology

A scenario-based approach to the design of human-computer interaction (Rosson and Carroll 2001, Carroll 2000) was adopted. The premise behind scenario-based methods is that descriptions of people using technology are essential in analyzing how technology is reshaping or will reshape their activities. A scenario is a story about people carrying out an activity[23]. Scenarios help the designer to understand technology as it will be experienced by "real" (or at least realistic) users, carrying out "real" activities in the context of their "real" work or play practices. They facilitate a more holistic approach to the design of technology: people's backgrounds, as well as the physical and social settings in which the technology will exist can be incorporated into the design process. Scenarios can be thought of as an inexpensive prototyping tool (Nielsen 1993, page 18).

The scenario-based design process begins with an *analysis of current practice* usually entailing some form of fieldwork. The findings from this analysis are used to write

---

[23] Erickson (1995) distinguishes stories from scenarios. Stories are less abstract, more detailed, and describe atypical situations. Erickson notes the value of using stories during the design process.

problem scenarios. A problem scenario is a story about the problem domain as it exists prior to the introduction of a certain technology. These problem scenarios are transformed into *activity scenarios*, which are narratives of typical services that users will seek from the system being designed. *Information scenarios* are elaborations of the activity scenarios which provide details of the information that the system will provide to the user. *Interaction scenarios* describe the details of user interaction and feedback. The final stage is prototyping based on the interaction scenarios and evaluation. The process as a whole from problem scenarios to prototype development is iterative.

Problem Scenarios and Interaction Scenarios

CoMem was designed using three sets of scenarios, all of which originate from the ethnographic study presented in Chapter 3. In the first set, a novice designer asks a mentor for help. This was developed into a scenario where the mentor is unavailable, and so the novice has to rely on CoMem to help him identify reusable items, and provide enough contextual information for him to be able to understand and reuse these items.

The second set of scenarios also starts with the novice asking his mentor for help. However, this was developed into a scenario where the mentor uses CoMem as a mentoring tool. He identifies parts of the corporate memory for the novice to explore on CoMem.

The third set of scenarios considers reuse from the perspective of an experienced designer, who is more concerned with productivity than understanding or learning. She knows what she is looking for, and uses CoMem to help her find it, retrieve it, and reuse it.

The iterative analysis and refinement of these scenarios were used to guide the development of CoMem.

The three scenarios are presented here only in the form of *problem scenarios* and *interaction scenarios*. The problem scenarios below express the problem situations addressed by CoMem as they exist in current practice and the interaction scenarios are the corresponding usage scenarios. Sample output of the scenario-based methodology is presented in Appendix A.

Novice Problem Scenario	Mentoring Problem Scenario	Expert Problem Scenario
An expert structural designer, Matthew, and a novice, Nick, both work for a structural design office in Northern California. The office is part of the "X Inc" structural engineering firm. They are working on a ten-storey hotel that has a large cooling tower unit. Nick must design the frame that will support this cooling tower. Nick gets stuck and asks Matthew for advice.		Eleanor is working on a staircase detail for an office building. This stair needs to start on a slab-on-grade and end on a composite slab. Eleanor has two options: she

Novice Problem Scenario	Mentoring Problem Scenario	Expert Problem Scenario

Matthew recalls several other hotel projects that were designed by "X Inc". He lists those to Nick and tells him that the Bay Saint Louis project, in particular, would be useful to look at.

Matthew walks with Nick to the room where old paper drawings are kept. Together they locate the set of drawings for the Bay Saint Louis project. Matthew takes out the structural drawings and briefly explains the structural system of the building to Nick. Matthew then finds the specific drawing sheet with the Bay Saint Louis cooling tower frame detail.

The drawing shows the cooling tower frame as it was finally built. It is a steel frame. Matthew realizes that what he had in mind for Nick to reuse is an earlier version that had a steel part and a concrete part. He is not sure if this earlier version is documented somewhere in the archive. Rather than go through the paper archive again, Matthew simply sketches the design for Nick. Matthew's sketch also shows the load path concept much more clearly than the CAD drawing would have, which helps Nick to understand the design. Matthew explains to Nick how and why the design evolved. Given the current project they are working on, it would be more appropriate to reuse the earlier composite version. Matthew recalls that the specifications of the cooling tower unit itself, which were provided by the HVAC (heating, ventilation and air conditioning) subcontractor, had a large impact on the design. Nick now feels confident enough to design the new cooling tower frame by reusing the same concepts as the Bay Saint Louis cooling tower frame, as well as some of the standard details.

can try to reuse a very similar stair which was designed by a colleague, or she can reuse a stair that she herself designed, but which was designed for a different situation. She decides that it would be easier to reuse her own stair because she can clearly remember the design process steps, and rationale for this stair, and so she can quickly make the necessary adjustments.

Her original stair was for a project called Woodside. She goes to her computer and opens the Woodside folder and the Drawings subfolder. She opens a file called 41760203.dwg. 417 and 602 are the project and sheet numbers respectively. She knows that 600 is the typical details category and she vaguely recalls that the stair detail was on the second sheet. 03 is the detail number. Eleanor does not see the detail she is looking for, so she opens 41760200.dwg, the entire sheet number 602 containing several details. The detail she is looking for is not there. She opens 41700100.dwg. This is the general notes sheet which contains a list of sheets. Eleanor realizes that she needed sheet 603, not 602. She opens sheet 603, finds the detail she needs, imports it to her current drawing, and makes the necessary changes.

From the above three problem scenarios, the following interaction scenarios were written which convey the envisioned usage situations of CoMem.

Novice Interaction Scenario

As before, Matthew and Nick are working on a ten-storey hotel that has a large cooling tower unit and Nick is assigned the task of designing the frame that will support this cooling tower. They are using the ProMem system. Nick gets stuck, but Matthew is not around to help. Nick clicks on the Reuse button in ProMem, which brings up CoMem. CoMem displays a map of the entire "X Inc" corporate memory. Items on the map are color-coded according to how relevant they are to his current project. Nick uses sliders to filter out irrelevant projects, disciplines, and components from the map. Most of the rectangles in the map are now grayed out. Of the few items that remain highlighted, Nick notices the Bay Saint Louis project. It has a relevant Engineering discipline, and several relevant components within that discipline. He clicks on the component labeled Cooling Tower Frame.

The project context and evolution history of the Bay Saint Louis cooling tower frame appear in two separate displays. Nick examines the evolution of the frame. He chooses to see only milestone versions of the evolution. He sees that it started as a

Mentoring Interaction Scenario

As before, Nick is assigned the task of designing the frame that will support a cooling tower for a large hotel project. Nick gets stuck and asks Matthew for advice. Matthew and Nick bring up CoMem together.

Matthew and Nick look at the Corporate Map. Matthew tells Nick to click on the rectangle labeled Bay Saint Louis Hotel. The project context and evolution history of the project appear onscreen. Nick clicks on the latest version in the evolution history and browses through the files and documents attached to this version. As he does this, Matthew describes the project informally from his memory. Matthew then instructs Nick to click on the structural system for this project. Nick finds the structural system in the Project Context Explorer and clicks on it. The Evolution History Explorer refreshes to display the evolution history of the structural system. As Nick browses through the versions of the structural system, Matthew starts to recount anecdotes from the project, his memory jogged by the notes and notifications displayed in the evolution

Expert Interaction Scenario

As before, Eleanor is working on the stairs for an office building. She opens CoMem, which shows the map of the "X Inc" corporate memory. Eleanor chooses to highlight items that she herself has worked on. She remembers that she has designed a set of stairs a few years ago for a project called Woodside. Eleanor already knows that her current project is completely different from Woodside, and the map confirms this: the rectangle on the map for Woodside is drawn in blue indicating low relevance. She also knows that the stair she designed for Woodside had a column in the middle. For her current project, she needs a cantilever stair.

Eleanor uses the keyword filter to highlight cantilever stairs. She notices a stair designed by a colleague that is given a high relevance rating by CoMem (it is colored bright red). Eleanor decides to investigate this stair further. She clicks on it to bring up its project context and evolution history.

Eleanor sees from the project context that the building had a concrete structure, like the current

45

composite steel-concrete frame but was later changed into a steel frame. He sees several notes that were exchanged between the architect and engineer that help to explain this change. Nick clicks on one of the versions, and a detailed view of this version appears. He finds a useful early sketch of the composite frame, which he saves to his local hard drive.

Next, Nick begins to explore the project context of the Bay Saint Louis frame. He clicks on the Engineering discipline object in the Project Context Explorer and sees that the Bay Saint Louis structural design criteria are similar to those in his current project. He notices a related component under the HVAC discipline: it is labeled Cooling Tower. This is the air conditioning unit that is supported by the frame. Nick finds a specifications sheet attached to this component. It gives him an idea of the loads for which he must now design his cooling tower frame.

history.

Matthew directs Nick to the Cooling Tower Frame component in the project context of the structural system. He tells Nick that this cooling tower frame can probably be reused in their current project. Nick clicks on the cooling tower frame to bring up its evolution history and Matthew, again his memory jogged by the evolution on screen, tells Nick an anecdote about the design of the frame. He describes the interactions he had with the project manager and architect that led to the frame being changed from a composite frame to a steel frame.

Matthew has to leave for a meeting. Having guided Nick to the Bay Saint Louis cooling tower frame, he feels confident that Nick can handle the design of the new cooling tower frame by reusing parts of the old frame. He reminds Nick to think critically about the old frame before reusing it, and reminds him that he will have to make several changes to adapt it to the current project. As he leaves, Matthew gives Nick one final instruction: that he (Nick) should document any improvements he makes to the design of the cooling tower so future reusers will benefit from his experience.

project on which she is working. However, this old stair detail starts and ends on a composite slab. For her current project, the stair needs to start on a slab-on-grade and end on a composite slab. This is a minor change to the detail. Eleanor goes to the evolution history of the stair, clicks on the latest version, and saves the CAD drawing to her computer.

Eleanor continues to search the Corporate Map for other stairs she can use. She expands her search to just the keyword *stair*. She sees several stairs from previous projects that show up as red rectangles on the map. However each one is unique and each is significantly different from the stair she needs to design for her current project. In several places, Eleanor sees stair components with no CAD graphics, but with a note saying that standard stairs will be provided by a stair manufacturer.

Research Methodology

Figure 9 shows an overview of the methodology adopted in this research. From the ethnographic study, it was observed that internal reuse is effective because the designer can find and understand the item he/she is reusing from his/her internal memory. In other words, relationship 1 in Figure 9 was empirically observed, i.e. the ability to find and understand internally leads to effective internal reuse. These ethnographic observations address the internal reuse aspects of the research questions listed in Chapter 1:

- How does *finding* occur in internal knowledge reuse?
- What is the nature of the project context exploration in internal knowledge reuse?
- What is the nature of the evolution history exploration in internal knowledge reuse?

Based on these observations, CoMem is designed specifically to support finding and understanding. This design process addresses the external reuse aspects of the research questions listed in Chapter 1: how can finding and understanding (through project context exploration and evolution history exploration) be supported in external knowledge reuse? Therefore relationship 2 in Figure 9 expresses the design rationale behind the design of CoMem. This is based on relationship 4 in Figure 9, which was posed as the hypothesis of this research.

The purpose of the formal evaluation described in Chapter 11 is to test relationships 3 and 4 in Figure 9, thereby testing the hypothesis of this research.

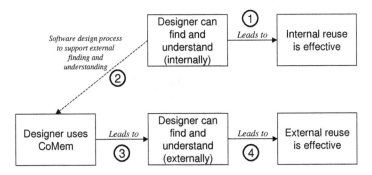

Figure 9: The research methodology.

Closing Remarks

AEC designers in current practice reuse designs mainly from their own personal experiences or by asking colleagues or mentors. In both cases, the process of *internal knowledge reuse* is central. Internal knowledge reuse is effective because the designer can *find* items to reuse from his/her internal memory, and can recall the context of these

47

items and can therefore *understand* them and reuse them effectively. A computer system for supporting reuse from an external repository should support *finding* and *understanding*. The external repository should emulate the characteristics of design knowledge as it occurs in the designer's internal memory. It should be a *corporate memory* of rich, detailed *knowledge in context*. These findings were used to design CoMem, a prototype corporate memory system, using a scenario-based design methodology. Three sample scenarios are presented here, in the form of problem scenarios and interaction scenarios, which correspond to three usage scenarios: mentoring, novice reuse, and expert reuse.

CoMem cannot replace mentoring. A human mentor will be more effective than CoMem in helping the novice to identify reusable items and to understand these items. In the presence of a mentor, CoMem's main role will be to allow the mentor and the novice cooperatively to retrieve externally encoded information such as electronic drawings or documents, or e-mails describing team interactions and design rationale, and to view all these disparate documents side by side. The mentor, using his/her own internal memory, will be able to weave all of these elements into a larger picture describing the project context and evolution history of the item.

CoMem must support reuse by a novice in the absence of the mentor. In this case, the mode of interaction will be closer to *exploration*. CoMem must generate a relevance measure that will guide the novice's exploration of the corporate memory and help him/her to find reusable items. Once the novice has found a potentially reusable item, CoMem must enable him/her to explore the context of the item in order to understand how and why it was designed the way it was.

Finally, CoMem must also support reuse by an expert designer who will probably be looking for a specific design from the corporate memory, as well as specific items of contextual information to help him/her adapt this design to the current project. Here the mode of interaction is closer to *retrieval*. The expert will rely less on the relevance measure and more on filtering tools that enable him/her to find items based on explicit search criteria.

Regarding the methodology for this research, the ethnographic observations detailed in Chapter 3 and summarized here address the internal knowledge reuse aspects of the research questions listed in Chapter 1 (how do finding and understanding occur in internal knowledge reuse). The CoMem design process outlined here and described in detail in the following four chapters address the external knowledge reuse aspects of the research questions listed in Chapter 1 (how can finding and understanding be supported in external knowledge reuse). The formal evaluation of CoMem described in Chapter 11 serves as a test of the hypothesis of this research.

48

Chapter 5

COMEM – A CORPORATE MEMORY COMPUTER ENVIRONMENT

This chapter gives a summary of CoMem, and the following three chapters examine each of CoMem's three main modules in more detail.

CoMem Modules for Supporting Reuse

The CoMem human-computer interaction experience is based on the principle of *"overview first, zoom and filter, and then details-on-demand"* (Shneiderman 1999). Based on the three reuse activities identified above – find, explore project context, explore evolution history – CoMem has three corresponding modules: an *Overview*, a *Project Context Explorer*, and an *Evolution History Explorer* (Figure 10).

Reuse step ⟶		User interaction
Find reusable item	"overview first, zoom and filter, and then details-on-demand"	Overview
Explore item's evolution history		Evolution history explorer
Explore item's project context		Project context explorer

Figure 10: CoMem HCI experience. Transformation from observed reuse steps to user interactions.

Figure 11 shows the views that are generated of the SME[24] corporate memory for each of the three modules. For each module, various metaphors were investigated, as well as possible visualization and interaction techniques (Fruchter and Demian 2002). *Metaphor* here is used in a human-computer interaction sense. Metaphors increase the usability of user interfaces by supporting understanding by analogy. Modern operating systems use the *desktop metaphor*. Online services use shopping cart and checkout metaphors to relate the novel experience of buying online to the familiar experience of buying at a bricks and mortar store. For a discussion of the advantages and pitfalls of using metaphors, see Nelson (1990).

---

[24] SME is the Semantic Modeling Engine, a schema for storing semantically annotated projects in the corporate memory (described in more detail in the 'Points of Departure' section on page 6).

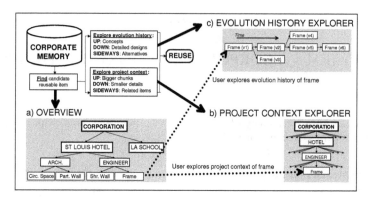

Figure 11: Views of SME data that are generated for each of the CoMem modules. (a) The Overview shows the entire corporate memory. (b) The Project Context Explorer takes a single item (in this case a structural frame) as its focal point. (c) The Evolution History Explorer shows the versions of a single item.

The Overview supports the designer in finding reusable items. The objective is to enable the designer to view the entire corporate memory at a glance. The Overview gives the designer an indication of which "regions" of the corporate memory contain potentially reusable items. The Overview might be extremely dense. Filtering tools are used to avert information overload and help the designer focus by adding emphasis to more relevant items. The design of the CoMem Overview module is described in Chapter 6.

Once the user has selected an item from the Overview, the *Project Context Explorer* supports the designer in exploring this item's project context. This module shows the project and discipline to which this item belongs, as well as related components and disciplines that would help the designer understand the found item. The item selected from the Overview becomes the *focal point* of the Project Context Explorer. The design of the CoMem Project Context Explorer module is described in Chapter 7.

In the third module, the *Evolution History Explorer*, the designer can explore the evolution history of any item selected from the Overview. This view tells the story of how this item evolved from an abstract idea to a fully designed and detailed physical artifact or component. The design of the CoMem Evolution History Explorer module is described in Chapter 8.

In addition to the Overview, Project Context Explorer, and Evolution History Explorer, CoMem also includes a *content viewer*, which displays all the disparate content associated with an item (text description, CAD file, hyperlinks, notes, notifications, data) in a single web page.

CoMem System Architecture

Figure 12 illustrates the CoMem system architecture. The Overview, Project Context Explorer, and Evolution History Explorer are implemented as a single Java application. The database, containing the accumulated set of project memories, has a C programming language interface. There is a Java class that has methods implemented in C using the Java Native Interface to connect to and retrieve data from the database.

The CoMem content viewer is implemented as a JSP web application.

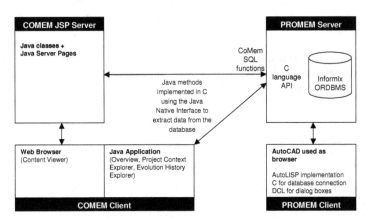

Figure 12: CoMem system architecture.

CoMem can run in an offline mode where it does not need to connect to the ProMem database server, but instead uses cached data saved using Java's serialization functions. A version of CoMem which runs in interactive workspace (Johanson et al. 2002) has been created, which allows the user to trigger CoMem modules to appear on any of the displays in an iRoom (Fruchter et al. 2007).

51

COMEM OVERVIEW MODULE[25]

Introduction

Internal reuse is effective because a designer can *find* reusable items from his/her internal memory. A reuse system for external reuse from a corporate memory must support *finding*. This chapter addresses the following question: *how can a corporate memory system support the designer in finding reusable items?*

This study argues that *finding* can be supported by providing an *overview* of the corporate memory which displays *all items* at a glance, and providing the user with filtering and navigation tools. This argument is based on ethnographic evidence (Chapter 3) and related research (Chapter 2).

The designer can identify potentially reusable items in the Overview and then explore the two contextual dimensions (project context and evolution history) of this item in separate modules.

Having made the case for an overview, this chapter explores how such an overview can be realized in CoMem. In particular, it describes the map interaction metaphor and the treemap visualization technique. Further possibilities in the design of treemaps are described, as well as how CoMem addresses them.

The Need for an Overview

An overview of large information spaces reduces search, allows the detection of overall patterns, and aids the user in choosing the next move (Card et al. 1999, Section 3.3). It is not obvious that an overview is the best approach for helping the user to find items from a large repository, especially if only a tiny fraction of the items shown on the overview are of interest to the user. For example, when submitting a query to a web search engine, it would be of little use to the user to see an overview of the entire World Wide Web.

Two characteristics of a corporate memory make it well suited to an overview. Firstly, it is a much smaller repository than the entire WWW, and so an overview is a realistic approach. Secondly, the corporate memory does not consist of a flat list of documents as most document repositories do. It is composed of a hierarchically structured collection of projects, building subsystems or disciplines, and individual components. It

---

[25] Some of the contents of this chapter were first published in Demian, P. and Fruchter, R., 2006, "Finding and Understanding Reusable Designs from Large Hierarchical Repositories." Information Visualization Journal, Volume 5, Number 1, pp. 28-46. They are reproduced here with permission of Palgrave Macmillan.

has been observed from the ethnographic study that the designer will need to make comparisons at all three levels of granularity simultaneously (Figure 13). This suggests the use of some visual overview that allows such comparisons to be made, rather than returning a flat list of "hits" that satisfy a query specified by the user.

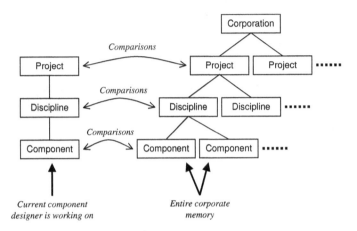

Figure 13: The designer makes comparisons at all three levels of granularity when finding reusable items.

An overview is in line with the interface design philosophy of CoMem, which is founded on achieving a balanced division of labor between the human and the computer based on their respective strengths. CoMem is based on the principle that the burden of finding reusable items should be left to the user as much as possible. In other words, CoMem provides the interaction mechanisms, but does not automatically identify reusable items. This is based on the observation that the human designer is much better equipped to make this assessment.

This principle can be put into practice by providing an overview that shows *everything* along with tools for filtering and zooming. CoMem generates a relevance measure to help the user identify reusable items, and this relevance measure (or any other search criteria specified by the user) can be displayed on the Overview.

Related Work on Overviews
The idea of *overview and detail*, i.e. providing an overview for orientation and one or more detailed views for further work, has been used for many years in information visualization systems. It is formally articulated by Card et al. (1999, Section 3.3).

The need for an overview is implied by *information foraging theory* (Pirolli and Card 1999). According to this theory, users of information will modify their information

53

seeking strategies or the structure of the environment to maximize their rate of gaining valuable information. These modifications are largely based on *information scent*. Information scent is the imperfect perception of the value, cost, or access path of information sources obtained from proximal cues, such as bibliographic citations, WWW links, or icons representing the sources. For an overview and detail interface, each item of detailed information must leave a sufficiently strong information scent in the overview in order to enable the information forager to make an informed decision about which piece of information to pursue. If there is no scent, the forager will perform a random walk.

Overviews have also been studied in the context of web site navigation. The objective of web navigation design is to enable the user to answer the following questions (Nielsen 2000):
1. Where am I?
2. Where have I been?
3. Where can I go?

For the purposes of CoMem, the third question is the most important. The symbols on the Overview must serve as sources of information scent, i.e. proximal cues to the chunks of information that they represent. The user's perception of this information scent enables him/her to make an informed decision about where to go next.

Constantly visible site maps (comparable to the idea of an overview being discussed here) were found to improve performance in information seeking tasks (Danielson 2002). They were found to provide a bird's eye view, reminding the user that there is a whole world out there waiting to be explored.

Users' web navigation habits are often characterized using a "hub and spoke" model (Catledge and Pitkow 1995). The overview can serve as a hub that provides links to all the items of information that are reachable.

Finally, some researchers have recognized the problems associated with showing everything on the overview, particularly for large information spaces. In addressing the question of whether the overview should show everything, Darken and Sibert (1993) write: "Putting more and more on a map is like jumping on a hot air balloon to get a good view of the city. The higher you go, the more you can see, but the increased altitude also decreases the strength of the stimuli."

The Map Metaphor
CoMem uses a map metaphor for the Overview. This metaphor emerged from the scenario-based design process as a useful way for thinking about the Overview. A map is traditionally defined as "a representation of things in space." Recent definitions have shifted the emphasis from strictly objective representations of physical space to more subjective representations that facilitate the spatial understanding of things, concepts,

conditions, processes, or events in the human world (Edson 2001). This more contemporary definition brings maps within the domain of *visualization*: the use of visual representations to amplify human cognition in support of particular tasks (Card et al. 1999, Chapter 1). Visualizations exploit human visual perception to amplify cognition to an extent that would be impossible using non-visual forms such as purely symbolic or textual representations. In particular, CoMem endeavors to exploit two properties of maps:

- The *spatial* property. A map is usually a smaller scale representation of an actual physical space. This small-scale representation effectively communicates the properties of *containment* and *proximity* between the entities represented on the map: Palo Alto is in California; Palo Alto is near Menlo Park.
- The *semantic* property. By overlaying certain marks (points, lines, areas, all represented according to some visual vocabulary) on the mapped space, a map is able to convey additional information (beyond containment and proximity) efficiently and rapidly. For example, political maps, topographic maps, natural resources etc.

A useful example is that of a weather map. The weather map, as a map of the local geography, is useful in its own right for someone unfamiliar with the area. Because the average newspaper reader is familiar with his/her local geography, simply by glancing at the weather map each morning, he/she can tell what the weather will be like in his/her area and the surrounding areas.

The Overview should express the "geography" of the corporate memory: which projects contain which disciplines and components, and which items are "close" to each other.

The CoMem user is expected to develop a familiarity with the geography of the corporate memory. Given a problem that he/she is working on, the map will appear with different areas highlighted to indicate that they are potentially reusable, and the user can tell at a glance which parts of the corporate memory to explore further. For novice users, areas of the map can be highlighted according to CoMem's measure of relevance to the users' current design problem. Expert users who do not wish to depend on CoMem's relevance measure can input their own queries, and the results from these queries are highlighted on the map. This is comparable to different information being superimposed on the map: weather, topography, political boundaries, resources, population density and so on.

The remainder of this chapter considers how to design this map of the corporate memory, or *Corporate Map*, so as to provide the maximum possible support for knowledge reuse[26].

---

[26] Modern writers on the history of cartography emphasize that is it impossible to study a map without considering its social context and the tasks for which it was intended. For example Harley (2001) rejects "cartographic positivism", the notion that cartography is objective, detached, neutral, and transparent. He denies that maps can be true or false, "except in the narrowest

Treemap Visualization

An SME corporate memory is a hierarchical data structure where a corporation contains multiple projects, a project consists of multiple disciplines, and a discipline contributes multiple components. This hierarchy can become very large (10^5 items)[27]. The Overview needs to show the entire corporate memory in a single display.

In a treemap visualization, projects, disciplines and individual components are represented as nested rectangles. The size of each rectangle is mapped to a measure of how much "knowledge" this node encapsulates. For example, an object that has a rich version history and is linked to many external documents and annotations will be assigned a larger area. The color of each rectangle is mapped to a measure of how relevant this object is to the designer's current design task. The advantages of treemaps are:

- They make full use of the available display space
- They complement the map metaphor[28].
- They are particularly effective for very large, fixed depth hierarchies, such as an SME corporate memory.
- If properly designed, they can support comparisons and assessment of relevance at all three levels of granularity simultaneously.

The classic treemap algorithm (Johnson and Shneiderman 1991) uses a slice-and-dice approach, subdividing each rectangle (representing a project, discipline, or component) either vertically or horizontally amongst a node's children (Figure 14). The most important disadvantage of the classic treemap is that its rectangles can have very high aspect ratios, which makes it difficult to select or label the rectangles, compare sizes, or perceive structural relationships between nodes.

Euclidean sense." In this sense, an accurate roadmap is not one that accurately depicts the roads, but one that will help a traveler to reach his/her destination. It is in this sense that this chapter talks about "designing" the Corporate Map.

[27] Consider a small corporation that has worked on 10 projects. Each project involves 10 disciplines or building subsystems, with each discipline contributing 50 components. If each object in the corporate memory was versioned 20 times over its lifetime, then the total number of items in the corporate memory is 10^5.

[28] In light of the above discussion of maps, the treemap maps the parent-child link relationship to enclosure in 2D space. It therefore conveys *containment*. It does not, as yet, convey *proximity* between similar siblings. The idea of mapping similarity between siblings to proximity on the treemap is explored on page 62. On a purely visual level, Fiore and Smith (2001) compare a treemap to a land-use map. They note that it is tempting to compare heavily subdivided rectangles to busy urban areas and large rectangles to calmer rural areas. However such a reading is flawed. The large rectangles represent not empty plains but vast leaf-nodes with huge amounts of data.

Figure 14: Generating a treemap using the classic treemap algorithm.

The squarified treemap algorithm (Bruls et al. 1999) and the clustered treemap algorithm (Wattenberg 1999) attempt to minimize the aspect ratios of the rectangles. Both algorithms produce very similar treemaps, although the squarified algorithm was found to give slightly lower aspect ratios (Shneiderman and Wattenberg 2001). Figure 15 shows a small corporate memory as a squarified treemap.

Figure 15: Generating a treemap using the squarified treemap algorithm.

The most important disadvantage of squarified and clustered treemaps is that changes in the sizes of the nodes produce dramatic changes in the layouts produced, which can be disorientating for the user. For example, the user might be used to seeing the architecture discipline of the Bay Saint Louis hotel always at the top left corner of the map. If this discipline increases slightly in size (for example because a new component is added to it) then it can suddenly jump to the lower right corner because this layout produces the most squarified rectangles. Such drastic layout changes can prevent the user from becoming familiar with the "geography" of the corporate memory.

Another algorithm, the ordered treemap (Shneiderman and Wattenberg 2001), addresses this issue. It attempts to maintain proximity relationships between nodes, which discourages large layout changes in dynamic data. The cost of this is slightly higher aspect ratios than those produced by the squarified or clustered algorithms. Ordering is discussed in more detail on page 62.

57

Treemap Design Issues

Emphasizing Structural Relationships
In most of the treemap applications presented in the literature, only the leaf nodes are of interest. Non-leaf nodes are important mainly for emphasizing structural relationships (i.e. grouping siblings together). Three techniques have been proposed for emphasizing structural relationships: framing (Johnson and Shneiderman 1991), padding (Turo and Johnson 1992, Fiore and Smith 2001, note that the former use the term *offsets*) and cushions (Bruls et al. 1999), as shown in Figure 16 (a), (b), and (c).

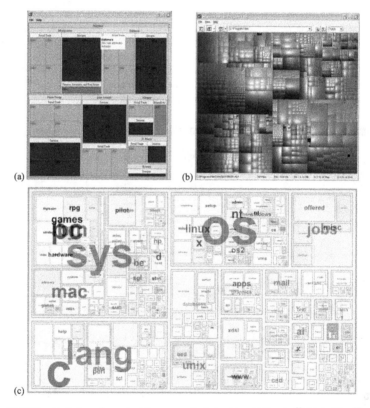

(a) (b) (c)

Figure 16: Different techniques for emphasizing structural relationships. (a) Framing[29]; (b) Cushions[30]; and (c) Padding[31].

---

[29] Screenshot from Treemap 3.0, developed by the Human Computer Interaction Laboratory at the University of Maryland.

58

For the purposes of this research, non-leaf nodes (projects and disciplines) must appear distinctly in the treemap so that they can be selected and explored independently of the components they contain. This is important for two reasons:

- A designer might want to reuse an item at the project or discipline levels of granularity. For example, the designer can reuse a document describing the structural design criteria that is linked to the structural discipline of the Bay Saint Louis project.
- Even if the designer is only interested in reusing components, he/she will need to assess whether a potentially reusable component comes from a similar discipline and project to the current design task.

CoMem uses padding, where gaps are left between a node's children, leaving the parent node visible behind the children. This enhances the map metaphor, giving the treemap the appearance of a contoured topographic map. Padding space is left around the perimeter of a set of siblings (Figure 17, top row), as well as between siblings (Figure 17, bottom row).

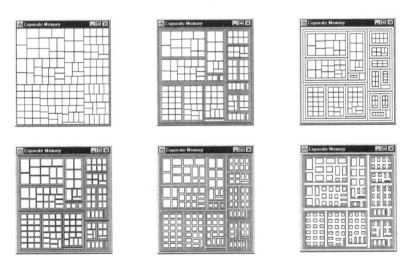

Figure 17: Increasing borders around sets of siblings (top row) and between siblings (bottom row).

Two further modifications are made to help emphasize structural relationships. Firstly, the amount of padding is increased with increasing level of granularity, so that sibling

---

[30] Screenshot from SequioaView, developed by the Department of Mathematics and Computer Science at the Technische Universiteit Eindhoven.

[31] Treemap produced by Fiore and Smith (2001).

projects have more padding space between them than sibling components. Secondly, the rectangle outline thickness is increased with increasing level of granularity, so that project rectangles are drawn with thicker lines than component rectangles.

Both of these measures eliminate the maze-like appearance of large treemaps, and help the designer to tell instantly whether the highlighted node is a project, discipline or component, and how relevant its ancestors and descendants are. However, these measures also reduce the density of the treemap (as a result of the padding, some smaller nodes are not drawn at all) and its accuracy (the padding distorts the relative sizes of the siblings). The resulting treemap is shown in Figure 18.

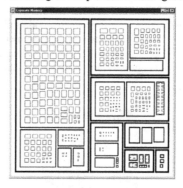

Figure 18: Treemap with varying padding and line thicknesses to help emphasize structural relationships.

Size Function
The size of each rectangle is mapped to a measure of how much content this node contains. For project and discipline objects, this size will be the sum of the sizes of the constituent component objects[32]. For a component object, the size can be a function of:

- The number of versions of this component
- The number of links to external documents
- The number of links to CAD objects
- The number of annotations (notes and notifications) attached to this component

---

[32] The treemap algorithm requires that the size of a node be greater than or equal to the sum of the sizes of its children. It would be possible to apply the same size function used for components to projects and disciplines. However, this size function will have to be designed carefully to ensure that this condition is always met. For the sake of simplicity, the size of a project or discipline object is taken as the sum of its children.

60

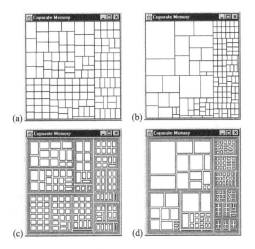

Figure 19: Experiments with the size function. Uniform size function (left column) versus exaggerated size function (right column); no padding (top row) versus padding (bottom row).

The screenshots in Figure 19 were produced with a simple size function where the size of a component is a function only of the number of times this component was versioned. Functions that produce a wide distribution of sizes make structural relationships more apparent in squarified or clustered treemaps. The difference can be seen by comparing treemaps (a) and (b) in Figure 19.

With padding it is not necessary to exaggerate the size distribution. In treemap (d) at the lower right of Figure 19, where the number of versions is raised to the power 2.5 to give a large size distribution, large nodes are disproportionately favored while smaller nodes are not drawn at all.

CoMem currently uses a simplified size function where the size of each node is directly proportional to the number of times it is versioned.

Color
The color of each rectangle is used to encode the relevance of this item to the designer's current design task, based on an automatically generated measure of relevance. Each node in the SME hierarchy will have an independent measure of relevance to the designer's current design task (see Chapter 10). This relevance measure is always in the range 0 to 1 [0,1] and is used to generate a color that is a linear interpolation between pure red (for relevant items) and pure blue (for irrelevant items). Combined with the padding which enables the designer to see the underlying rectangles for projects and disciplines, this coloring is extremely effective for making comparisons at all three levels of granularity simultaneously.

61

Ordering

The squarified algorithm lays out siblings in order of decreasing size. This generally leads to rectangles with smaller aspect ratios. However, this results in an arbitrary placement with regards to similarity. In keeping with the map metaphor, it is desired that "similar" siblings be laid out closer to each other. This would result in similar subsets of siblings forming meaningful "regions" on the map which, when the relevance measure is indicated on the map, would appear as patches of high or low relevance. The user can therefore explore relevant regions more closely. Because similar nodes are laid out closer together, the user is more likely to find relevant items serendipitously while exploring a nearby node.

One possibility is to use the ordered treemap algorithm, ordering the nodes by their relevance to the current design task. This would produce relevant regions as desired. However, the generated layout will depend on the designer's current design task (from which the relevance measures are calculated). This means that the designer will see wildly varying layouts depending on the current design task he/she is working on. The aim is for the colors only, and not the treemap layout, to depend on the current design task. A better approach would be to order components within a discipline by *class*[33], so that components of the same class are laid out near each other.

Another possibility is to pre-compute an affinity matrix for each set of siblings that expresses how similar a node is to each of its siblings. This affinity matrix can then be used to lay out similar nodes closer to each other. There is currently no algorithm that lays out a treemap using an affinity matrix.

Currently, CoMem uses the squarified treemap algorithm, and so variations in layout as the corporate memory grows and evolves remain a potential problem.

Labels

Each rectangle on the treemap serves as an "information scent" leading to the item it represents. The size and color of the rectangle are important components of this information scent ("how much content will I find there, and how relevant is this content?"). Text labels can significantly increase this information scent by jogging the user's memory, particularly for retrieval tasks where the user already has some idea what he/she is looking for.

Labeling treemaps is particularly challenging. CoMem centers each label horizontally over its rectangle. Vertically, the labels are positioned either one third of the height from the top or one third of the height from the bottom, alternating between the two for adjacent rectangles. This improves the distinctiveness of each label; with the labels

---

[33] In SME, each discipline has a set of classes, created and modified by the designer as the project progresses. For a structure discipline, the set of classes may include *beam, column, frame*, and so on. Each component within a discipline is assigned to one of the classes. The component is in effect an instantiation of the class.

simply centered vertically, the labels for a row of rectangles were found to resemble a continuous line of text.

The labels can either be scaled to fit the rectangle, or fixed sizes can be used, with project labels being the largest and component labels the smallest. CoMem currently provides both options. Figure 20 shows the labeling options available in CoMem. The advantages and disadvantages of each approach are listed in Table 3. Generally speaking, the scaled labels are more effective. If the label is scaled down below a threshold value, it is not drawn at all. In the case of the fixed size labels, choosing the label font sizes is extremely difficult because the rectangles vary widely in size within each level of granularity.

With both scaled and fixed size labels, occlusion of underlying labels has proven to be problematic. To address this, CoMem can draw partially transparent labels. The transparency is increased with increasing level of granularity (i.e. project labels are almost completely transparent and the component labels are completely opaque). The rationale behind this is that project object labels are more likely to be large and therefore occlude other labels. However this is not always the case, particularly for disciplines or components with very short labels that become large when scaled up to fit the rectangle. In such cases, the label of a discipline can be larger than that of its project, and the fact that the discipline label is more opaque can be confusing. Another possibility is to assign the transparency of each label based on its size rather than level of granularity.

Table 3: Advantages and disadvantages of various labeling options.

	Advantages	Disadvantages
Scaled labels	The labels are as large as space permits, and so they are generally more legible. The user can easily associate the label with its rectangle (because the label extends from one end of the rectangle to the other). The size of the label functions as a convenient criterion for choosing which labels to draw: if after scaling the label will be too small to be legible, it is not drawn at all.	Short labels, when scaled up, are disproportionately prominent. For example, a discipline label can appear more prominently than the project to which it belongs because the discipline has a short name.
Fixed size labels	It is easy to identify the level of granularity of an item from the font size of its label.	It is almost impossible to select suitable font sizes to use for the whole treemap, because the rectangle sizes vary widely within each level of granularity. If the rectangle is too small, the label can overflow outside the two sides of the rectangle. If the rectangle is too large, the label is drowned by empty space. In both cases, it becomes hard to associate the label with its rectangle. There are no obvious criteria for choosing which labels to draw (given that with *all* labels drawn, occlusion renders the whole treemap almost illegible). One possibility is relevance.

One simple refinement of scaled labels that addresses the problem of short labels appearing disproportionately large is to enforce the rule that no item is to have a larger label than its parent. A further refinement is to draw discipline labels in a different color so that the user can tell whether a label refers to a project or a discipline (Figure 21).

Further research and design are needed to improve the labels in CoMem. As a supplement to the labels painted on the treemap, CoMem also displays the description of each rectangle in the form of a "tooltip". If the user briefly lingers with the mouse pointer over a rectangle, then the description appears in a small box (Figure 22). Therefore, the text description is available even if the rectangle is not labeled, or if the label is too small or is occluded.

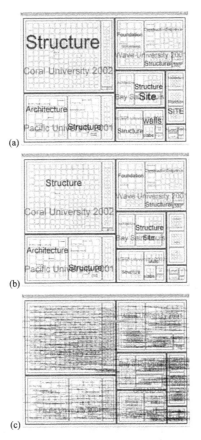

Figure 20: Labeling treemaps. (a) Labels are scaled to fit the rectangle; (b) labels are scaled to fit the rectangle but cannot exceed the size of the parent label; (c) fixed sizes of labels are used for project, discipline, and component labels, in which case all labels are drawn.

Figure 21: A labeled Corporate Map in which discipline labels are green, and project and component labels are yellow.

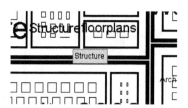

Figure 22: A tooltip appears when the user lingers with the mouse pointer over any rectangle.

Filtering

At the beginning of this chapter, it was argued that the Overview needs to show the *entire* corporate memory. However, even for a small corporate memory, the Overview can be extremely dense to the extent that the user is unable to distinguish or click on individual rectangles. It will be necessary to allow the designer to add emphasis to certain parts of the corporate memory that are more relevant. There are two possible interaction mechanisms for adding emphasis to items on the treemap:

- Filtering out undesired items using dynamic querying
- Zooming in on potentially reusable regions of the map

CoMem currently allows the user to filter out items using *dynamic querying*. In a dynamic querying environment, search results are instantly updated as the user adjusts sliders or selects buttons to query a database (Shneiderman 1994). A designer can filter out items based on:

- **Relevance**. The user can filter out projects, disciplines, or components that have a small relevance measure. (The relevance measure is based on a text analysis of the items in the corporate memory and the current design task.)
- **Date**. The user can filter out items that were begun after a certain date or completed before a certain date.

66

- **Keywords**. These can be applied separately at the component, discipline and project levels, i.e. at each level of granularity. For example, the user might only be interested in *hotel* projects with *atriums*, in the *structural* disciplines from these projects, and particularly in *cooling tower frames*.
- **Ownership**. Each item has a set of people associated with it, who contributed to its design over the course of its evolution history. Designers frequently want to limit their search to items that they themselves have worked on, or that a specific person has contributed to.

Filtered items can be grayed out, allowing the user to focus on the remaining brightly colored items. Alternatively, filtered out items can be omitted, leaving more space for the remaining items. For a large corporate memory, it will probably be necessary to filter out some items in this way in order to make the remainder of the items on the map discernable. Figure 23 illustrates filtering in CoMem.

The second possibility for adding emphasis to potentially reusable items is to zoom in on these regions of the map. This would be more consistent with the map metaphor. A zoomable version of the Corporate Map is currently being developed.

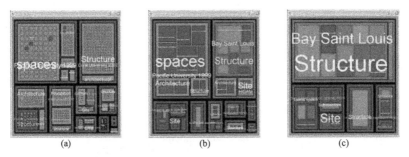

Figure 23: Filtering in CoMem. (a) Filtered items are grayed out. (b) Filtered components are not drawn at all. Filtered projects or disciplines are drawn grayed out if they have unfiltered components, otherwise they are not drawn at all. (c) Filtered components are not drawn. Filtered projects and disciplines are "pruned", i.e. they are not drawn, regardless of whether or not they have unfiltered children.

Figure 24 shows the control panel used to apply filters to the CoMem Overview.

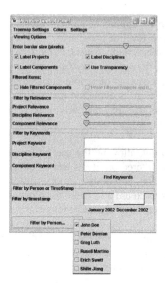

Figure 24: The CoMem Overview Control Panel, which gives the user various coloring and labeling options, and allows the user to filter by relevance, keywords, timestamp, or person.

Closing Remarks

How can the user of a corporate memory system be supported in finding reusable items, particularly if the repository is large? It was argued that the user needs to see an *overview* of the entire corporate memory. The user can identify potentially reusable items in the overview and then explore the two contextual dimensions (project context and evolution history) of this item in separate modules.

CoMem uses the *Corporate Map* for the Overview, where the projects, disciplines, and components in the corporate memory are visualized as nested rectangles using the *squarified treemap algorithm*. Treemaps are an effective technique for visualizing large hierarchies such as an SME corporate memory. The Corporate Map provides a succinct overview at a glance of the "geography" of the corporate memory: which projects contain which disciplines and components. Over time, the user should develop a familiarity with the Corporate Map.

The color of each rectangle is used to encode the relevance of that item to the designer's current design task (i.e. the component or discipline on which the designer is currently working). This visual indication of relevance, combined with the user's familiarity with the geography of the corporate memory, should enable the user to quickly identify relevant regions to explore at greater depth.

68

Varying the treemap padding and line thickness are used as means of emphasizing structural relationships within the treemap. If the user notices a relevant item on the map, these measures should enable the user to tell instantly whether this item is a project, discipline, or component, and how relevant its ancestors and/or descendants are. The objective is to support reuse and comparison at all three levels of granularity simultaneously.

Filtering is described as a mechanism for adding emphasis to items that are more likely to be reusable and averting information overload. Future research will investigate the use of zooming, which is more in line with the map metaphor.

The CoMem evaluation (Chapter 11) investigates the claims made in this chapter. The Corporate Map will be compared to traditional interfaces to test whether it provides improved support for finding tasks (both exploration and retrieval).

Chapter 7

THE COMEM PROJECT CONTEXT EXPLORER[34]

Once the designer has identified a potentially reusable item from the CoMem Overview (the Corporate Map), he/she needs to explore the project context of this item. This *focal item*, i.e. the building subsystem or component being reused, was not designed in isolation but as part of a larger project. The focal item needs to be considered in its project context if it is to be understood and successfully reused.

The first constituent of the focal item's project context consists of *its ancestors and descendants* in the hierarchy. For example, a braced frame is part of a larger structural system, which in turn, is part of a whole building. The braced frame consists of subparts: beams, columns, and connections. In the *six degrees of exploration,* this is referred to as upward and downward exploration of the project context (see Chapter 3, page 26).

The second constituent of the project context consists of *related items* in the parts hierarchy that lie outside of the path to the focal item and its sub-tree. A building consists of many intricately interrelated subsystems and components. The braced frame may be embedded in an architectural partition wall, or may be designed to be extra strong because it supports a library on the floor above. In the *six degrees of exploration* this is referred to as sideways exploration of the project context.

This chapter examines how a user can interact with a focal item's project context in order to understand that item and reuse it effectively. In particular, two problems need to be addressed:

- Firstly, how can *related items* be identified? Whereas the ancestors and descendents emerge naturally from the structure of the data, the *relatedness* between the focal node and its related items needs to be inferred.
- Secondly, how can the focal item be represented with its ancestors, descendents, and related items so as to support upward, downward, and sideways exploration? This is both a visualization problem and an interaction design problem.

The Fisheye Lens Metaphor and the Fisheye View
CoMem uses a fisheye lens metaphor for the Project Context Explorer. This metaphor was suggested by Furnas (1981) as part of his *fisheye view.* In contrast to a zoom lens,

---

[34] Some of the contents of this chapter were first published in Demian, P. and Fruchter, R., 2006, "Finding and Understanding Reusable Designs from Large Hierarchical Repositories." Information Visualization Journal, Volume 5, Number 1, pp. 28-46. They are reproduced here with permission of Palgrave Macmillan.

which provides local detail at the expense of the global view, a fisheye lens simultaneously combines local detail with global context.

The *fisheye view* (Furnas 1981) gives a methodology for generating a small display of a large information structure by controlling the field of vision, in analogy to a fisheye lens. Given a focal point, Furnas defines a *degree of interest function* over the remaining data items. Given a focal point, the user will not be equally interested in all items. Furnas decomposes the degree of interest into *a priori* and *a posteriori* components.

The *a priori* component is a contribution to an item's degree of interest which transcends the given interaction, but depends on the global importance of that item. The *a posteriori* component is the contribution to an item's degree of interest that depends on the current focal point, and is derived from some measure of distance between that item and the focal point.

Furnas goes on to describe the special case of fisheye views of hierarchical tree structures. In a hierarchical tree structure, the *a priori* component can be taken as the *level of detail* of an item (i.e. how high up the hierarchy it is). For an SME corporate memory, this maps to the level of granularity of each item. A project object is intrinsically more important than a discipline object, which is intrinsically more important than a component object. The *a posteriori* component can be taken as the *distance* to the focal node (i.e. the number of links in the shortest path between the focal node and the node in question).

Therefore, for a hierarchy such as an SME corporate memory, one possible formulation is as follows:
1. Focal point: '.'
   This is the focal item selected from the CoMem Overview, whose project context the user is exploring.
2. The *a posteriori* component of the degree of interest of node x is the distance between the focal point and x:
   $$d(x,.)$$
   This is the number of links on the path between node x and the focal point.
3. The *a priori* component of the degree of interest is the Level Of Granularity:
   $$LOG(x)$$
   For a tree structure, this is defined as:
   $$-d(x,r)$$
   This is the distance between node x and the root r of the tree. Therefore:
   If x is the corporation, $LOG(x) = 0$
   If x is a project object, $LOG(x) = -1$
   If x is a discipline object, $LOG(x) = -2$
   If x is a component object, $LOG(x) = -3$
4. The Degree Of Interest of node x is:

71

$$DOI(x|.) = LOG(x) - d(x,.)$$
$$DOI(x|.) = - d(x,r) - d(x,.)$$

Applying these equations to a small sample corporate memory gives the values shown in Figure 25.

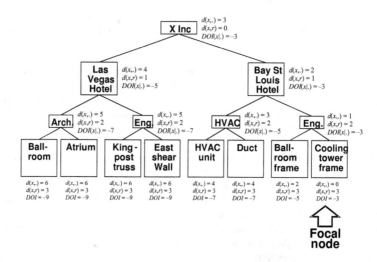

Figure 25: Degree of interest values for a small hierarchical corporate memory given the specified focal node.

An interesting property of the formulation above is the emergence of what Furnas calls *iso-interest contours*, those points on the tree with the same degree of interest. This can be visualized by "picking up" the tree from the root and the focal node, and letting the remaining nodes dangle below (Figure 26).

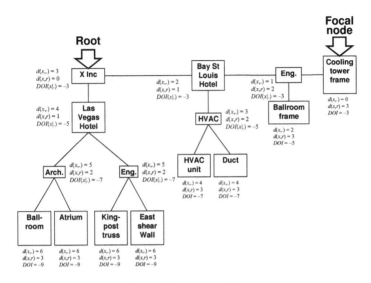

Figure 26: Iso-interest contours.

The fisheye view as formulated above for tree structures is useful because it addresses the first problem noted above: how to identify related items given a focal item or node. It can be argued that items with a higher degree of interest are more closely related to the focal node, and are more likely to help the user understand the focal node. However this formulation by itself is not sufficient to effectively identify related items because it is based only on structural relationships within the tree and does not take into account the contents of each node. For example in Figure 26 above, both the *Las Vegas Hotel Architectural* subsystem and the *Las Vegas Hotel Engineering* subsystem are assigned the same degree of interest when compared to the *cooling tower frame* component. By common sense, the *Engineering* subsystem is more closely related to the *cooling tower frame* component, because the *cooling tower frame* is itself part of an *Engineering* subsystem from the *Bay St Louis Hotel* project. In any case, the fisheye view provides a useful starting point and possible extensions to it are discussed below.

Node-Link Diagrams of Fisheye Views
The fisheye view can be used to define a degree of interest function over a set of nodes in a hierarchy given a focal node. The issue of how to visualize such a hierarchy with varying degree of interest still remains to be addressed. Ideally, items with a higher degree of interest should be displayed more prominently.

There are two categories of techniques for visualizing hierarchies:
- techniques using *connection* (i.e. node-link diagrams)

73

- techniques using *enclosure* (i.e. treemaps)

CoMem uses a traditional node-link diagram to visualize the project context of a given item. Items are laid out in a 2D space where the horizontal axis is the degree of interest (exploiting the iso-interest contours) and the vertical axis is the level of granularity.

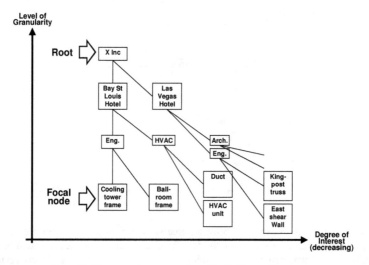

Figure 27: Diagrammatic representation of the CoMem project context.

This assignment of the axes is consistent with the six degrees of exploration formalized above; the designer moves *up* to explore this item at a coarser level of granularity, *down* to look at finer grains, and *sideways* to explore related items. Figure 27 shows this diagrammatically, and Figure 28 shows an actual screenshot from CoMem.

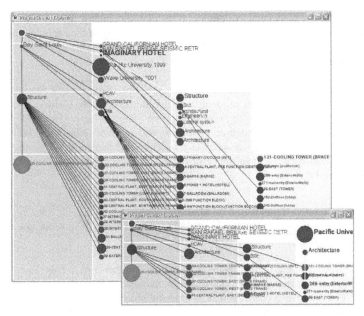

Figure 28: Screenshots of the CoMem Project Context Explorer. The window is resizable. Each object is positioned in the vertical axis according to its level of granularity, and in the horizontal axis according to its degree of interest with respect to the focal point.

As with the Corporate Map, the color of each item is used to denote this item's relevance and the size is used to denote the volume of content attached to this item. The relevance measure is calculated between the focal item and every other item in the corporate memory[35].

Essentially, each rectangular area (quadrant) in the Project Context Explorer contains a category of relatives. In Figure 28, the focal item is a component, *36-Cooling Tower (Brace Frame)*, and it is drawn in red at the bottom left grey quadrant. The next quadrant to the right contains the focal item's siblings. The subsequent quadrant to the right contains the focal item's first cousins, followed by the second cousins.

Within each category of relatives, the items are sorted by relevance to the focal item, so that more relevant items appear closer to the top of the list. The display is resizable, and once space has run out, the remaining items on the list are not painted. If an item's parent is painted, then a black line is painted between the parent and the child.

---

[35] This relevance measure is calculated in the same way as that for the Overview module (see Chapter 10). Whereas in the Overview each item was compared to the designer's "problem item", here each item is compared to the focal item.

In order to support contextual exploration, if the mouse pointer is moved over any item, then that item and all of its ancestors are highlighted. If any of the ancestors have been pruned because of lack of space, they are temporarily painted in the white area at the top right of the display.

This addresses the concern noted above that the simple fisheye view formulation for trees by itself is not sufficient to identify related documents. In CoMem, the fisheye degree of interest is used as the *primary* measure of *relatedness*. However, between sets of items with the same degree of interest and level of granularity, the CoMem relevance measure is used as a *secondary* measure of relatedness. This measure of relevance is also used, in the absence of unlimited display space, to prune all but the most closely related items[36].

Identifying Related Items Based on Shared Graphics
We now return to the question of identifying related items to discuss one possible refinement to the approach taken by CoMem.

In SME, a CAD object can belong to multiple components. This is crucial for facilitating communication and coordination in multidisciplinary teams using shared 3D models. For example, a *partition wall* component object created by the architect and a *shear wall* component object created by the engineer can both be linked to the same graphic object. Even though they are semantically two distinct objects, they are physically the same building component. At a later time, when this *partition wall* is being reused, the *shear wall* object will be an important part of its project context. It will provide valuable information about this component from a structural perspective.

The presence of shared graphics between two items is an important way of inferring that these items are closely related. Figure 29 shows a small sample corporate memory where a graphic object is shared amongst several components. There are three levels of *relatedness* that can be inferred from shared graphics. In decreasing order these are:
- Graphics shared within the same discipline. This might occur if the same graphic object is part of two or more interacting components within the same building subsystem. For example, the same room can serve as both a classroom and conference room within the same architectural subsystem of a university project.
- Graphics shared within the same project. As noted, one of the strengths of SME is that it allows graphic objects to be interpreted differently by team members working on different building subsystems within the same project.
- Graphics shared from project to project. For example, a standard stair detail which is repeatedly reused from project to project. This is not currently possible within the ProMem system. It is included for completeness. In the future, if

---

[36] This use of the CoMem relevance as a secondary measure of relatedness can be considered a refinement of the *a posteriori* component of interest within a set of siblings. It is also conceivable to make use of *a priori* refinements to help with pruning siblings, for example by volume of knowledge.

CoMem tracks instances of reuse between projects, then this information can be used to enrich the project context. This idea of *knowledge refinement* is outside the scope of this research.

These levels of relatedness through shared graphics can be used by CoMem as a refinement to the *a posteriori* component of the degree of interest. They can be used in the same way as the relevance measure to prune out items if space is limited. Alternatively, items that share graphics with the focal item can be visually highlighted[37].

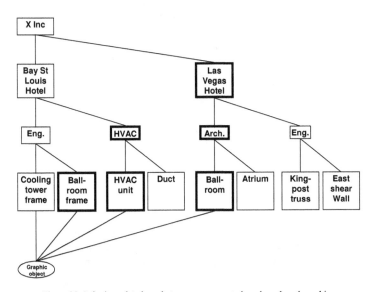

Figure 29: Inferring relatedness between components based on shared graphics.

Focus+Context Visualizations

We will now turn to the question of how to visualize and interact with a focal item and its project context. Several alternatives that were considered when designing CoMem will be discussed.

Furnas' fisheye view, described above, places more emphasis on *how much* context will be displayed rather than *how* it will be displayed. In its simplest applications, the user specifies a cutoff degree of interest value, and items with interest below that value are not displayed at all. Other techniques place greater emphasis on *visually* combining

---

[37] This refinement has not been implemented in CoMem because it is a computationally intensive operation that would drastically slow down the system.

local detail with global context. These are referred to as focus+context techniques. Focus+context techniques address the second problem identified above: how to visualize and interact with a tree structure by combining a detailed view of a particular node with a view of its context. However for the most part, these techniques do not address the first problem: how to identify related items. For most focus+context techniques, "context" refers to the whole tree rather than a subset of related items.

Cone trees (Robertson et al. 1991) visualize trees in 3D, allowing much bushier trees to be displayed (Figure 30). Nodes near the front of the 3D space are considered to be the focal points, and nodes near the back are the context. Furnas' fisheye formulation has been applied to cone trees to reduce the number of nodes on the screen.

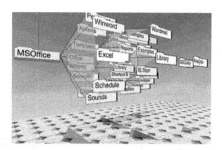

Figure 30: A cone tree[38].

In the hyperbolic tree (Lamping and Rao 1995), the nodes in a hierarchy are positioned in hyperbolic rather than Euclidean space. Any node can be dragged into the center of the hyperbolic plane thereby bringing it into focus, while keeping the entire hierarchy visible.

Figure 31 shows a series of hyperbolic trees for the *Bay Saint Louis* project from the "X Inc" corporate memory. In Figure 31 (a) the entire hierarchy is shown. In Figure 31 (b) the *structural engineering* discipline is dragged to the center. The designer can see that the project also included *site*, *HVAC* and *architecture* disciplines, but the components belonging to those disciplines are pruned out to keep the display simple. Finally in Figure 31 (c) the *cooling tower frame* is dragged to the center of the display. In this view the ancestors of the *cooling tower frame* component (the *structural engineering* discipline and *Bay Saint Louis* project) are not clearly visible.

---

[38] Screenshot from User Interface Research at Palo Alto Research Center.

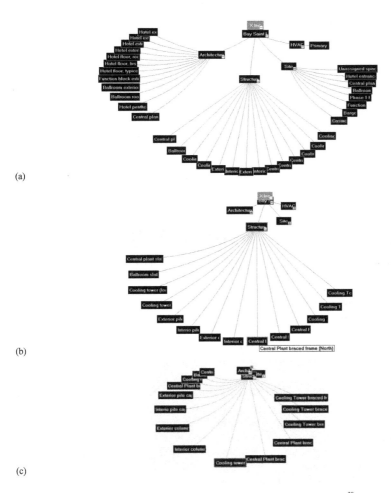

(a)

(b)

(c)

Figure 31: A series of hyperbolic trees for the cooling tower scenario[39].

One possibility that was explored is the use of an outline tree[40]. The outline tree can be transformed into a focus+context view by mapping the degree of interest to the size of the icon and to the text used to label this item (Figure 32). This alleviates the problem of scrolling in an expanded tree. As the user changes the focal node, the size of each

---

[39] These screenshots were produced using Inxight Tree Studio.

[40] The term *outline tree* will be used to describe Microsoft Explorer-style interfaces where hierarchies are visualized using indented lists of icons and labels that can be collapsed or expanded.

item is updated. The user is still able to expand and collapse trees and sub-trees in the usual way.

Figure 32: A fisheye outline tree view of the corporate memory. The *NW Function Block* is the focal node.

The focus+context techniques described above were explored but eventually rejected for use in CoMem. They were found to add little value beyond the simple node-link visualization used by CoMem.

Treemaps Revisited
Treemaps were used to provide an overview of the entire corporate memory. Can treemaps also be used to explore the project context of a focal node? One possibility is to color each rectangle by its degree of interest value (calculated using the fisheye formulation described above). Figure 33 shows a treemap of the cooling tower scenario colored by degree of interest. This visualization is not as effective as the fisheye view in Figure 28 because it does not emphasize the focal node and degree of interest distribution as much as the node-link diagram. In addition, it would be confusing to use the same interaction design for the two separate tasks of finding and understanding.

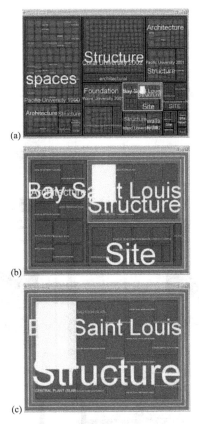

Figure 33: A treemap colored by degree of interest relative to the Bay Saint Louis cooling tower. The set of screenshots shows a series of interactions in which the user filters progressively more based on degree of interest.

In Figure 33, the user gets a closer view of the focal node (the highlighted and solid yellow rectangle) by pruning out items with degree of interest lower than a cutoff value, and gradually increasing that value. This can be referred to as *fisheye zooming*.

A better approach might be to combine enclosure with *spatial* (rather than fisheye) zooming. *Zoomable user interfaces* (for example Perlin and Fox 1993) have been shown to be more effective than their non-zooming counterparts for many applications. These applications include image browsing (Combs and Bederson 1999[41]) and web browsing (Bederson et al. 1996).

---

[41] This study found that the zoomable image browser was only marginally better than a standard 2D browser.

81

Figure 34 shows a series of screenshots depicting a typical interaction where the user starts with a view of the entire corporate memory and progressively zooms in to a specific component (the focal node). All the time, the user can see the discipline and project to which the component belongs, as well as related components, disciplines, and projects in the corporate memory.

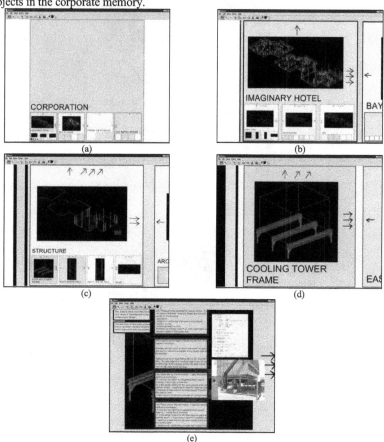

Figure 34: Exploring the project context using zooming. A series of screenshots depicting a typical interaction where the user starts with a view of the entire corporate memory and progressively zooms in to a specific component[42].

The advantages of this approach are:

---

[42] These images were prepared using the Jazz Java Toolkit (Bederson et al. 2000).

- The ability to zoom means that screen real estate is almost unlimited, with lots of space to display CAD drawings, sketches, and documents linked to the SME objects.
- The SME hierarchy can be laid out in a self-similar manner, so that the interaction is the same at every level of the tree. A similar view is generated when a sub-tree is magnified. Components, disciplines, and projects are all displayed in a similar way[43].

There are two major disadvantages to this approach. Firstly, the global view of the corporation as a whole and of the current project is lost as soon as the user zooms in. Secondly, it does not emphasize the degree of interest of each item. Recall that in the node-link diagrams of Figure 27 and Figure 28, the items were positioned on the horizontal axis according to their degree of interest. In Figure 33, items are colored according to their degree of interest.

Closing Remarks
The objective in the fisheye view is to enable the designer to explore the project context of a given item in the corporate memory. Recall that this exploration can be *upwards* (exploring ancestors: disciplines and projects), *downwards* (exploring descendents: components and CAD objects), or *sideways* (exploring related items).

The fisheye view formulation is presented here as a formal mechanism for assigning a *degree of interest* to each item in the corporate memory given a focal node. The project context is then visualized by laying out the hierarchy in a 2D space where the horizontal axis is the degree of interest and the vertical axis is the level of granularity. In addition, a relevance measure is generated between each item and the focal item. This relevance is denoted using the color of each node and is used to prune less relevant nodes among nodes with the same degree of interest if space is limited. Table 4 contrasts this approach with the other approaches mentioned in this chapter.

---

[43] This self-similarity property has been called the *fractal tree layout* (Koike and Yoshihara 1993).

Table 4: A comparison of the approaches considered in this chapter.

Approach	How are related items identified?	How is the project context visualized?	How does the user interact with the project context?
CoMem	Fisheye formulation for tree structures. CoMem relevance measure is used to order and if necessary prune nodes among sets of items with the same degree of interest and level of granularity.	The hierarchy is laid out in 2D space where the horizontal axis is the degree of interest and the vertical axis is the level of granularity. The color of each node denotes its relevance (compared to the focal node) and the size denotes the volume of content attached to it. Items that share graphics with the focal item can be visually highlighted.	The user highlights a node by moving that mouse pointer over that node. The node and its ancestors are highlighted using a thick yellow outline. If any of the ancestors had been pruned, they are temporarily painted at the top right of the display.
Cone trees	For both Cone trees and Hyperbolic trees, related items are assumed to be nearby in the tree, and so will be visible when the focal node is brought into focus. Fisheye formulation can be used to prune nodes.	The hierarchy is laid out in 3D space, with a set of children forming a cone below their parent.	The user brings the desired node into focus by dragging it to the font of the display.
Hyperbolic trees		The hierarchy is laid out in hyperbolic space. Nodes near the focal node are displayed more prominently near the middle of the space. Nodes further away from the focal node are displayed less prominently at the periphery.	The user brings the desired node into focus by dragging it to the middle of the display. Non-leaf nodes can be expanded or collapsed.
Fisheye Outline Tree	Fisheye formulation for tree structures.	The hierarchy is represented as a standard expandable/collapsible tree. The color of each node denotes its relevance (compared to the focal node) and the size denotes its fisheye degree of interest.	By expanding and collapsing sub-trees.

Approach	How are related items identified?	How is the project context visualized?	How does the user interact with the project context?
Treemaps colored by degree of interest	Fisheye formulation for tree structures.	Treemap (nested rectangles). The color of each rectangle is mapped to the fisheye degree of interest.	The user can filter out items by degree of interest.
Zoomable treemap	Fisheye formulation for tree structures.	Nested rectangles are laid out on a zoomable canvas. The color of each rectangle is mapped to the fisheye degree of interest.	The user clicks on an item and the camera zooms in spatially onto this item.

The main advantage of the CoMem approach is that the interaction maps directly to the three degrees of project context exploration (up, down, and sideways). More than any of the other approaches mentioned, CoMem emphasizes the degree of interest (by using it to position the nodes) to help focus the user's exploration efforts. By exploiting the iso-interest contours, the resulting layout of the hierarchy highlights structural relationships surrounding the focal item. At the same time, using the relevance measure to color and prune nodes if necessary serves to highlight related items that are not necessarily structurally close to the focal item[44].

The stated objective of displaying high degree of interest nodes more prominently is achieved because a large number of low degree of interest nodes share the same amount of space as that shared by the relatively small number of high degree of interest nodes. Furthermore, relevant nodes (according to CoMem's relevance measure) that are buried deep in the tree, and would have been otherwise difficult to find, are always displayed prominently at the top of the list.

The main disadvantage of CoMem's Project Context Explorer is that it is not as interactive as other approaches. The subset of contextual nodes that are displayed is a function of the space available (i.e. the size of the window) and the user cannot interactively choose to show more, less, or different nodes.

Cone trees and hyperbolic trees address this by effectively visualizing the entire hierarchy in a limited space. Their major disadvantage is their implicit assumption that related items will be near the focal node (in terms of number of links). Related items that are not near the focal node are not prominently displayed.

---

[44] This dichotomy between providing *structural* and *associative* links is noted by Nielsen (1999) in the context of web design. Associative links connect information chunks based solely on content similarity and relevance. Structural links connect information chunks based on the global structure of the web site. A well-designed web site needs both types of links.

The fisheye outline tree attempts to alleviate the problem of scrolling in outline trees by using the fisheye degree of interest to assign less space to nodes with smaller interest. In theory, a fully expanded tree can be displayed in a single screen. In practice, this would require an unreasonable amount of reduction in the size of items with less interest. If the reduction is limited to keep all labels legible, then the user will either have to scroll or collapse some sub-trees. The fisheye outline tree will still depend on the user exploring the project context by scrolling or expanding sub-trees to find related items deep in the hierarchy.

The treemap and zoomable treemap both abandon *connection* for visualizing hierarchies in favor of *enclosure* (Card et al. 1999, Section 2.4). However treemaps tend to obscure structural relationships which, while less important in the CoMem Overview, are crucial when exploring the project context. The second problem with treemaps is that a choice must be made between mapping the color of each rectangle to the fisheye degree of interest or to the CoMem relevance measure. However, as noted above, it is the combination of the two that is quite powerful. If the treemap is colored by fisheye degree of interest, upward and downward exploration is supported (particularly by filtering) but sideways exploration of the project context becomes ineffective.

The zoomable treemap was found to add little value. Its main advantage is its almost unlimited space which allows the content (graphics, notes, images, documents) attached to each item to be displayed on the same zoomable canvas rather than in a separate display. Its main disadvantage is that it is not really a fisheye view: the user has to choose between a global or local view.

Chapter 8

COMEM EVOLUTION HISTORY EXPLORER

In the Overview of the corporate memory the designer can find potentially reusable items. In the Project Context Explorer he/she can explore the project context of this item. However in both of these views, the time dimension is "flattened".

The Evolution History Explorer should enable the designer to explore the evolution history of a given project, discipline, or component over time. The exploration can be upward (exploring early concepts), downward (exploring detailed designs), or sideways (exploring design alternatives). This is important for two reasons:

- **Reusing intermediate versions.** The reusable knowledge may be at an intermediate stage of the evolution of the item. For example, a fully designed CAD model of a cooling tower frame may not be useful, whereas an early sketch showing the load path concept is. Perhaps an early design alternative that was abandoned for the original project can now be reused.

- **Understanding a particular version and gaining design expertise.** Even if the final design can potentially be reused, the evolution of this design needs to be studied in order to *understand* this item and make an informed decision about whether and how to reuse it. It may be the case that the *process* is more important than the *product*. Importing a CAD component from a previous project will bring about an immediate improvement in productivity. However this can be small compared to the lasting improvement in productivity which results from understanding the design rationale and, as a result, *gaining valuable design expertise* to be applied to future projects.

These two objectives serve as yardsticks against which any solution for supporting evolution history exploration can be assessed.

The extent to which these objectives are accomplished depends not only on the user interface but also on the nature of the evolution history data available. The evolution history for an SME object is a tree structure. Each time the ProMem system detects a change in the design, a new version (node) of the object is created and linked to its parent. Each version has attached content:

- The specific *graphic objects* from the product model to which this semantic object is linked.

- *Notes* and *data objects*, which are attached to the product model in the same way that designers in current practice annotate paper drawings with handwritten notes.

- *Notifications objects*, which are used to solicit feedback, give approval, broadcast changes, or initiate negotiations in the same way that designers use e-mail during the design process. Notifications would also act as a substitute for requests for information (RFIs) during the construction phase.

- *Hyperlink objects*, which are used to share documents with team members in the same way that designers in current practice e-mail and fax documents to each other.

Collectively, the content attached to these versions describes both the nature of the evolution (how the design evolved) as well as the rationale for this evolution (why it evolved the way it did).

This chapter looks into how such a version history can be visualized and how the user can interact with it in order to offer the maximum possible support for the above objectives, in particular the second (understanding a particular version and gaining design expertise), which is more challenging.

The Storytelling Metaphor

CoMem uses a *storytelling* metaphor for the Evolution History Explorer. In its most literal meaning, a *story* is simply a narrative of facts or events. However stories have additional expressive content, weaving details, characters, and events into a whole that is greater than the sum of its parts (Simmons 2001). In a traditional Jewish allegory[45], *Truth is turned away from every door in the village because her nakedness frightens the people. When Parable finds her huddled in a corner, she has pity on her and takes her home. There, Parable dresses Truth in story and sends her out again. Clothed in story, Truth once again knocks on the villagers' doors, and this time is readily welcomed into their houses.*

The use of stories has been studied in a wide variety of contexts, including influencing people, particularly in business settings (Simmons 2001), bringing about social change (Davis 2002), and as a literary art form (for example Fulford 1999).

Storytelling is a useful metaphor for two different but related reasons. Firstly, storytelling is how expertise is usually transferred in professional practice, and secondly, design rationale in ProMem is captured in the form of a story.

The first reason is more important because it relates directly to the user. Storytelling is a useful metaphor because it matches how young designers interact with "human corporate memories", i.e. the experienced designers and mentors at design practices. Our ethnographic observations of designers at work show that experienced designers tell stories. When instructing novice designers on how to reuse a component from a previous project, they tell stories about how this component was originally designed. Even when answering more general questions about how to solve certain design

---

[45] Recounted in Simmons 2001.

88

problems, experienced designers refer back to specific projects from their past experiences and recount anecdotes from those projects.

The second reason has to do with the nature of knowledge capture in ProMem. Knowledge capture in ProMem is *process-based*. Design rationale is captured as a *history of the design process*. Put simply, ProMem captures *the story* of how a team of designers got together and designed a building. It cannot be said to capture design expertise in any formal way (such as by the formulation, application, or refinement of rules), but the design expertise possessed by the team members is manifest in the story of their collaboration.

An exhaustive account of ProMem's approach for capturing design rationale is beyond the scope of this discussion (Fruchter et al. 1998). Briefly, Regli et al. (2000) contrast process-based approaches with *feature-based* approaches. Feature-based approaches capture design rationale as a series of logical moves within a precisely defined design space. Process-based approaches are useful when the problems are vague, there is little or no standardization of the designed artifact, and the design process is supported rather than automated. Feature-based approaches are useful for task specific contexts and narrow design domains where the domain knowledge can be formally encoded. Multi-disciplinary building design falls into the former category. Almost no two buildings are the same, nor can the domain knowledge from the ten or so different disciplines that contribute to the design of a building be exhaustively codified. Instead, ProMem captures design rationale by supporting typical communication and coordination tasks that occur in building design teams. These include annotating the building model with notes, sharing data or documents linked to the building model, and sending change notifications to solicit feedback, give approval, broadcast changes, or initiate negotiations. Each time a change is detected (for example due to the addition of a note or a change in the CAD model), the system automatically creates a new version of the objects in question. This approach results in a relatively informal description of how the design evolved, but minimizes the additional effort required for knowledge capture.

In spite of its informality, the SME design evolution history is extremely valuable. Schön (1983) notes that expertise (particularly in design) lies not in rules or plans entertained in the mind prior to action, but in the action itself. Before him, Polanyi (1966) coined the term *tacit knowing* to describe the fact that "we know more than we can tell" – that knowledge which is capable of shaping behavior and yet is not ordinarily accessible to consciousness and so is difficult to capture directly.

More recently, researchers are beginning to recognize that *design is a social process* (Leifer 1997), and that design expertise lies not only in the individual designer's actions, but also in the interactions within a design team. Ferguson (1992, page 32) writes, "Those who observe the process of engineering design observe that it is not a totally formal affair, that drawings and specifications come into existence as a result of a social process. The various members of a design group can be expected to have divergent views of the most desirable way to accomplish the design they are working on… informal negotiations, discussions, laughter, gossip, and banter among members of a design group often have a leavening effect on the outcome." Arias et al. (1997) observe that "each stakeholder [in a design team] has a (sometimes narrow) view of the problem and an agenda to satisfy his/her particular goals. Stakeholders are often unaware that achieving their own goals can make things worse for other stakeholders." Bucciarelli (1994) concurs, proposing a model of the design process where each participant operates within a different "object world"[46].

If then design expertise cannot always be reduced to rules or procedures but is "in the design action" itself and much of what constitutes design action is the communication that goes on within a design team, then the story of how a design emerged from the communication within a design team can be said to capture to a large extent the designers' design rationale.

These two reasons – (1) design expertise is transferred in practice using stories, and (2) the design rationale is captured by ProMem in the form of a story – have in common their shared sense of a story as a conduit of knowledge. What formal reasoning fails to grasp, a story simply conveys[47]. The storytelling metaphor therefore expresses to the user that he/she is interacting with a narrative of the evolution of a designed component, and that this narrative is useful in its own right, just as a story told by a mentor is useful.

Visualizing Version Hierarchies in CoMem
Gershon and Page (2001) explore the link between storytelling and visualization. They propose two techniques: animation and frame-by-frame storytelling (the "comic book metaphor"). Garcia et al. (2002) use animation to communicate multidisciplinary design perspectives by adopting cinema storytelling techniques.

CoMem adopts the frame-by-frame technique, which gives the user more control over which parts of the story to explore and enables him/her to compare multiple versions side-by-side. The evolution history for an SME object is a tree structure, and so the story is not linear but may involve several design alternatives being explored in parallel. This tree has a very small average branching factor, usually around 1-3[48]. CoMem

---

[46] Interestingly, Bucciarelli (1994) proposes "story making" as a useful metaphor for the process by which *each participant* understands the designed artifact. However, this is different from my sense of a story of how the design emerges through a *collaborative* process. Bucciarelli later goes on to note that the various participants must "bring their stories into coherence" (page 84).

[47] Schank (1990) proposes a model of intelligent behavior based on creating stories from experiences, and then storing, retrieving, and telling these stories. "Intelligence depends upon the ability to translate descriptions of new events into labels that help in the retrieval of prior events. One can't be said to know something if one can't find it in memory when it is needed. Finding a relevant past experience that will help make sense of a new experience is at the core of intelligent behavior."

[48] Project evolution histories were studied in detail as part of the ProMem project (Reiner and Fruchter 2000). It was found that project teams often start with a consensus version of the shared 3D model, from which each member creates an individual private version. At a later time, the team members meet to resolve conflicts, and the individual private versions are merged

90

retrieves the versions of any item and any attached content from the database (Figure 12 in Chapter 5) and visualizes the version history using a node-link diagram[49], where each node is a version in evolution history and a "frame" (or *panel*, to use comic book terminology) in the story.

The versions are laid out on a *canvas*. Each version is represented as a color-coded circle. The color of the outline of the circle denotes its *level of importance* flag (*low*, *conflict*, or *milestone*), and the color of the center of the circle denotes its *level of sharing* flag (*private*, *public*, or *consensus*)[50]. Any content linked to this version is also displayed as an icon linked to the circle. The user double-clicks on the icon to see a full view of the content. If the content is a piece of text (a note, change notification, or piece of data) or an image, the full view is inserted into the canvas. For external documents that cannot be displayed on the canvas, double-clicking on the icon opens that document in an external window using the appropriate application.

The user is able to interact with this story in three ways. Firstly, the user is able to pan and zoom around the canvas. Secondly, the user is able to directly manipulate individual items on the canvas to move them or scale them. Thirdly, the user is able to filter out versions based on their levels of importance or sharing.

Figure 35 to Figure 39 illustrate a typical series of interactions with the Evolution History Explorer.

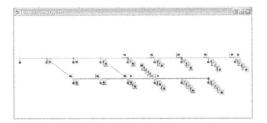

Figure 35: The CoMem Evolution History Explorer.

into a consensus version once again. According to this model, the version history is not a tree structure because a consensus version can have more than one parent. The current implementation of ProMem makes the simplification of requiring each version to have only one parent.

[49] A treemap was rejected in favor of a node-link diagram for two reasons. Firstly, unlike the SME object hierarchy, a version tree does not "imply" enclosure. Drawing the versions as nested rectangles does not offer a natural representation. Secondly, the version history has a linear, temporal element to it, as it tells a story unfolding over time. Treemaps encourage lateral, all-at-once readings, while node-link diagrams encourage linear, successive readings particularly if the tree is deep and the branching factor is small.

[50] This information is provided by the original design team working in ProMem. They are able to go back and flag various versions according to their level of importance and level of sharing.

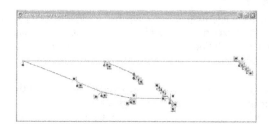

Figure 36: The user filters out unimportant versions.

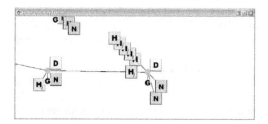

Figure 37: The user zooms in on one version.

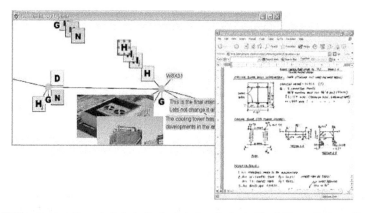

Figure 38: The user views the content attached to the desired version by double-clicking on the icons. Texts and images are displayed on the canvas; binary files are opened in the appropriate application in separate windows.

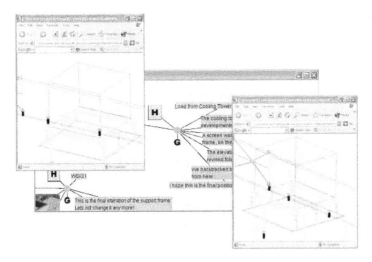

Figure 39: The user compares two different versions side-by-side.

Discussion and Closing Remarks

At the beginning of this chapter, two objectives for the Evolution History Explorer were laid down. These are that it should enable the user (1) to view and reuse intermediate versions, and (2) to *understand* the particular version being reused and *learn* from the expertise of the original designers by seeing their rationale.

The CoMem Evolution History Explorer clearly allows the user to see intermediate versions of the design, and so the first objective is accomplished. The degree to which the second objective is supported depends on how much content the original designers attached to their shared product model. It is assumed, for the sake of discussion, that the corporate memory is fairly richly annotated such that:

- most of the annotations that designers would normally make on paper drawings are included in the database in the form of note and data objects; and
- most of the files and documents that would normally be exchanged amongst team members by fax or e-mail are included in the database as hyperlink objects.

Given that this data is in place, how does the CoMem Evolution History Explorer help the user to understand the particular version being reused and gain valuable design expertise? The strength of the Evolution History Explorer is that it enables the user to see, interact with, and *therefore compare* multiple versions simultaneously. The user can consider each version as an episode in a larger story rather than as an isolated event. These comparisons can be made *vertically* (i.e. up and down exploration of the evolution history, Figure 40) or *horizontally* (i.e. sideways exploration of the evolution

93

history, Figure 41)[51]. Such comparisons would not be possible if the versions were displayed in a list, or if the user could only see one version at time (Figure 40).

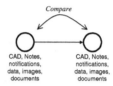

Figure 40: Vertical comparisons: comparing successive versions of any item from the corporate memory.

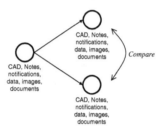

Figure 41: Horizontal comparisons: comparing alternatives of any item from the corporate memory.

Such comparisons are possible because the user is able to directly manipulate two or more versions and place them side-by-side. The user can then pan and zoom to obtain a good view of those two versions, and expand some of the content attached to them and examine it. It is possible to view a lot of content; because of zooming, the space available is virtually unlimited. The combination of the three functions above (moving and scaling individual items, panning and zooming on the canvas, filtering items based on flags) give the user the complete freedom to generate the perfect view and make comparisons in order to understand the story.

ProMem captures versions automatically each time even the smallest change is made to the design. As a result, the number of versions can be very large, with insignificant changes between consecutive versions. This might not reflect the user's idea of a version. For this reason, the ability to filter by the flags specified by the original design team is extremely useful. If the team flagged two versions as milestone versions, then these versions were probably meaningful milestones with important design

---

[51] The terms *vertically* and *horizontally* are used to be consistent with the six degrees of exploration presented in Chapter 3 (upwards or downwards exploration of versions earlier or later in time is vertical exploration and sideways exploration of alternatives is horizontal exploration), even though the versions are visualized in reverse orientation in both the figures in this chapter as well as in the CoMem Evolution History Explorer.

94

developments and changes occurring between them. The filters enable the user to make more meaningful comparisons by ignoring insignificant versions and preventing information overload (Figure 42). This filtering can either exclude intermediate versions in between milestone versions in the case of vertical comparisons (Figure 42 (a)) or insignificant versions in two or more parallel design alternatives in the case of horizontal comparisons (Figure 42 (b)).

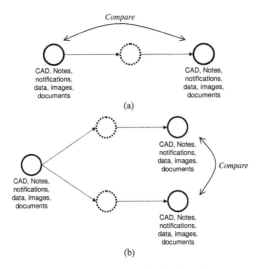

Figure 42: Comparisons against filtered versions.

Vertical comparisons enable the user to observe the differences between consecutive versions reflected in changes in the CAD objects and the attached content (Figure 40 and Figure 42 (a)). The horizontal comparisons enable the user to directly compare alternatives considered by the original design team (Figure 41 and Figure 42 (b)). As the original designers made decisions, their rationale is recorded in the content: the notes and notifications they exchanged and the information from external documents such as vendor catalogs or results from analysis and simulation programs. These act as snippets of rationale, with the user filling in the gaps as much as possible.

Emerging theories of comic book rhetoric (Duncan 1999, McCloud 1993) provide clues as to the effectiveness of this visual storytelling, particularly the concept of *encapsulation*: the framing of essential moments of a story in significant images. In the case of comic books, the creator makes the decision of what moments of the story to present. In CoMem, the user and the original designers jointly play this role. The original designers flag versions according to their levels of importance and sharing, and the user can filter according to these flags.

Duncan (1999) notes that encapsulation is a reductive process, i.e. the creators reduce the story to moments on a page by encapsulation. Readers expand the isolated moments represented in discrete panels into a continuous story by *closure*, the process by which they "fill in the gaps". The placement of panels side-by-side is essential for this process of closure.

To summarize, the CoMem Evolution History Explorer visually lays out the versions from the evolution history onto a canvas, with each version linked to its parent (Figure 43). The user is able to filter out unimportant versions and modify the initial arrangement by translating and scaling the elements so that important versions are positioned close to each other. Seeing the versions side-by-side facilitates the making of comparisons akin to the process of closure by which the reader of a comic book reconstructs a story from a series of discrete moments[52].

Because the user is able to "see" the story in this way, he/she will be able to explore and understand the rationale of the original designers in a way that would not be possible if the evolution history were presented as a flat list of versions or as a set of static images using a flipbook metaphor. A formal evaluation of this claim is presented in Chapter 11.

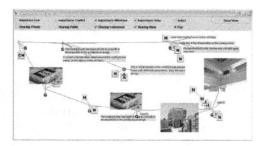

Figure 43: An example of the value of seeing different versions side-by-side in the CoMem Evolution History Explorer.

---

[52] Terry and Mynatt (2002) also note the importance of seeing different versions simultaneously side-by-side. In their case, this is important for supporting rapid and fluid experimentation and exploring of alternatives during creative work.

Chapter 9

A COMEM USAGE SCENARIO

In the three preceding chapters, the three modules of CoMem were each described in detail. In this chapter, CoMem usage is illustrated by revisiting the interaction scenario from Chapter 4, where Matthew and Nick are working on a ten-storey hotel that has a large cooling tower unit and Nick is assigned the task of designing the frame that will support this cooling tower. They are using the ProMem system (Figure 44). Nick gets stuck, but Matthew is not around to help. Nick clicks on the Reuse button in ProMem, which brings up CoMem (Figure 45). CoMem displays a map of the entire "X Inc" corporate memory. Items on the map are color-coded according to how relevant they are to his current project. Nick uses sliders to filter out irrelevant projects, disciplines, and components from the map (Figure 46). Most of the rectangles in the map are now grayed out. Of the few items that remain highlighted, Nick notices the Bay Saint Louis project. It has a relevant Engineering discipline, and several relevant components within that discipline. He clicks on the component labeled *Cooling Tower Frame*.

The project context and evolution history of the Bay Saint Louis cooling tower frame appear in two separate displays (Figure 47). Nick examines the evolution of the frame. He chooses to see only milestone versions of the evolution (Figure 48). He sees that it started as a composite steel-concrete frame but was later changed into a steel frame. He sees several notes that were exchanged between the architect and engineer that help to explain this change. Nick clicks on one of the versions, and a detailed view of this version appears (Figure 49). He finds a useful early sketch of the composite frame, which he saves to his local hard drive.

Next, Nick begins to explore the project context of the Bay Saint Louis frame. He clicks on the Engineering discipline object in the Project Context Explorer and sees that the Bay Saint Louis structural design criteria are similar to those in his current project (Figure 50). He notices a related component under the HVAC discipline: it is labeled Cooling Tower. This is the air conditioning unit that is supported by the frame. Nick finds a specifications sheet attached to this component (Figure 51). It gives him an idea of the loads for which he must now design his cooling tower frame.

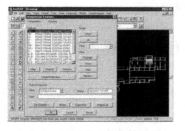

Figure 44: Nick is working in the ProMem system when he gets stuck. He presses the *Reuse* button.

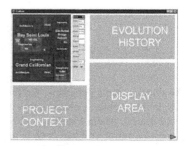

Figure 45: CoMem pops up on the screen and displays the Corporate Map.

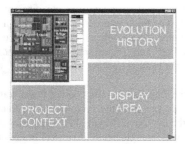

Figure 46: Nick filters out some items from the map using the sliders. He notices the cooling tower frame and clicks on it.

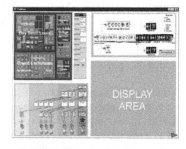

Figure 47: The project context and evolution history of the cooling tower are displayed.

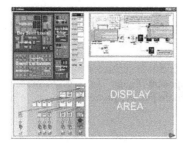

Figure 48: Nick filters out unimportant versions from the cooling tower evolution using the slider and enlarges several thumbnails.

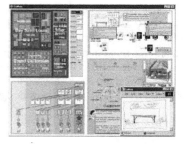

Figure 49: Nick clicks on a particular version from the Evolution History Explorer. The details of this version appear in the display area.

 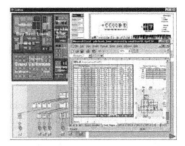

Figure 50: Nick uses the Project Context Explorer to view information about the structural system. He clicks on the document icon to bring up the design criteria document in a separate window.

Figure 51: Nick uses the Project Context Explorer to view information about the cooling tower unit. He views a spreadsheet attached as a hyperlink to the cooling tower object.

Figure 44 to Figure 51 express how the CoMem modules would be used in a typical interaction scenario. An evaluation of CoMem, specifically its relevance measure and usability, will be explored next.

Chapter 10

MEASURING RELEVANCE IN COMEM[53]

Introduction

Measuring relevance is an important task in CoMem. It is used for both the Overview and the Project Context Explorer.

In the CoMem Overview, a relevance measure is generated between the item the designer is working on when he/she brings up CoMem and every other item in the corporate memory. This relevance is used to color each rectangle on the CoMem Overview. This is important because the user, particularly if he/she is a novice who is unfamiliar with the contents of the corporate memory, might not know what to look for and might be unable to formulate an explicit query such as a keyword search. CoMem therefore uses the designer's current design task (the item he/she is working on) as an *implicit query* (Ye and Fischer 2002), and uses the results from this implicit query to highlight potentially reusable items on the map.

In the CoMem Project Context Explorer, a relevance measure is generated between the *focal item* and every other item in the corporate memory. The relevance measure is therefore used to identify related items that must also be explored by the designer to help him/her understand the focal item. In this case, the relevance measure augments the *degree of interest* which is based purely on structural relationships within the corporate memory hierarchy.

This chapter addresses the following question: how can the relevance between any two corporate memory objects (be they project, discipline, or component objects) be measured? Based on the tasks that this relevance measure is intended to support, *relevance* can be defined as follows:

For any two corporate memory objects A and B, object B is relevant to object A if:

- The designer is currently working on object A and object B is potentially reusable. For example, if the designer is working on the *cooling tower* for the *Bay Saint Louis* hotel, then the *cooling tower* from the *Las Vegas* hotel is relevant because it is potentially reusable. Or:
- The designer is considering reusing object A and object B is somehow related to object A, such that knowledge about object B helps the designer to understand object A. For example, if the designer is considering reusing a spiral staircase from a previous project, then the structural system supporting that staircase is

---

[53] Some of the contents of this chapter were first published in Demian P. and Fruchter R., 2005, "Measuring relevance in support of design reuse from archives of building product models", ASCE Journal of Computing in Civil Engineering, volume 29, issue 2, pp. 119-136. They are reproduced here with permission from ASCE.

relevant because that structure had an important impact on the design of the staircase and will help the designer understand and therefore effectively reuse that staircase.

The general approach taken by CoMem in generating relevance measures is to use text analysis. This is effective because SME objects are annotated with text strings that represent the meaning of CAD objects from a particular design perspective but have otherwise little formal data. A typical semantic annotation of a CAD object is usually one or two terms, such as *pile foundation*. On the other hand, this is challenging because those text strings are *much shorter* than those normally used for text analysis and retrieval. In information retrieval applications, documents are typically *at least* 200 words long. This chapter will therefore proceed as follows:

First, it is considered how CoMem objects can be converted into texts or *documents*. Having converted each object into a document, the result is a collection of documents. Techniques from the field of information retrieval will be considered for the comparison of documents.

Next, it is explored whether the performance of the relevance measure can be improved by considering structural relationships within the CoMem hierarchy. CoMem documents are not part of a flat collection but actually belong in a structured hierarchy. An item's ancestors, descendants, and possibly other "relatives" can be taken into account when making comparisons involving that item. This chapter concludes with a discussion of the approaches considered and suggestions for future research.

Converting Objects to Documents
CoMem objects contain textual information that is used by the designers of these objects to label, annotate, and collaboratively discuss each object[54]. The nature of this textual information will depend on whether the object is a project, discipline, or component.

Each project object has a name (for example *Bay Saint Louis* in Figure 52). This becomes the text of the project document. In theory, projects may also have note and notification objects linked to them, but the current implementation of ProMem does not support this.

Each discipline has a name and list of classes that constitute the vocabulary or ontology of that particular perspective on the design. The user is free to use any text string as a class; there is no universal vocabulary from which the class list is drawn. In theory,

---

[54] CoMem objects also contain non-textual information in the form of links to graphic objects from a shared product model and links to binary files such as sketches or calculations. An extension to the research presented here would be to attempt to include this non-textual information in the assessment of relevance.

101

disciplines may also have note and notification objects linked to them, but the current implementation of ProMem does not support this.

Each component object has a name, and belongs to (is an instance of) one of the classes in its parent discipline object. In addition, a component object may have one or more note, notification, graphic, or data objects linked to it.

Figure 52 gives examples of typical CoMem objects. Each object is converted into a document by concatenating all of its text elements as shown in Table 5. Separate examples are shown for component objects with and without note objects linked to them.

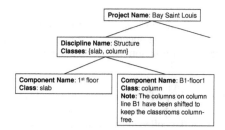

Figure 52: Typical CoMem objects. Each project object has a name and is linked to multiple discipline objects. Each discipline object has a name and a list of classes, and is linked to multiple components. Each component object has a name and belongs to a class.

Table 5: CoMem objects are converted to documents by concatenating all of the textual elements of the object.

Object Type	Object Text	Document Text
Project	Project Name: Bay Saint Louis	bay saint louis
Discipline	Discipline Name: Structure Classes: {slab, column}	structure slab column
Component (without note)	Component Name: 1st floor Class: slab	1st floor slab
Component (with note)	Component Name: B1-floor1 Class: column Note: The columns on column line B1 have been shifted to keep the classrooms column-free.	b1 floor1 column the columns on column line b1 have been shifted to keep the classrooms column free

Evaluating the Measures of Relevance

The tests described in this chapter were conducted on a pilot corporate memory consisting of 10 project objects, 35 discipline objects, and 1036 components, giving a total of 1081 objects (before versioning) that were converted into documents. The objects were fairly similar to the examples given in Table 5. Of the 1036 component objects, approximately 30% were annotated with note and change notification objects.

Common words that add little meaning (*the, you*, etc.) were identified using a list of about 400 common words and were filtered out prior to indexing. Otherwise, all terms were indexed[55]. Most of the experiments were repeated with and without stemming. Stemming is the process whereby distinct terms are reduced to their common grammatical root. Where stemming was used, it was performed using Porter's algorithm (Baeza-Yates and Ribeiro-Neto 1999, Appendix). For example, *concreting using concrete* would be indexed as *concret us concret*.

Evaluation of relevance measures was carried out in the classic information retrieval manner. Given a set of queries, and a set of documents for each query judged to be relevant by a human expert, the results returned by the relevance measure were compared to those provided by the expert using measurements of recall versus

---

[55] Because of computational limitations, it is sometimes desirable to reduce the number of terms in the index. One common approach is to index only those terms that occur above some threshold frequency in a document or in the collection as a whole. This approach was rejected because we also run the analysis with concatenated documents, which would have caused more terms to be indexed. This would make it difficult to make meaningful comparisons between the two sets of runs (with and without concatenated documents). Therefore, we simply index all terms that do not appear in the list of common words.

103

precision. Recall is the proportion of documents deemed *relevant* that were retrieved by the tool. However, this measurement does not take into account the case of *irrelevant* documents being retrieved alongside the relevant documents. Hence precision is used to calculate the proportion of retrieved documents that are actually *relevant*. For example, documents A, B, and C are deemed relevant; the tool retrieves C and D. Recall is 1/3 and precision is 1/2.

In the case of CoMem, a query is a specific object from the corporate memory, and the "hits" returned are other objects that are relevant to the query object. For each query, precision was measured at 11 standard recall levels from 0 to 1.0 in increments of 0.1 using the interpolation rule described in Section 3.2.1 of Baeza-Yates and Ribeiro-Neto (1999). The precision measurements at those recall levels were averaged for entire sets of queries.

Three sets of queries were considered independently: queries where the query object was a project (8 queries), a discipline (18 queries), and a component (6 queries). Those were considered separately because those three types of documents differ significantly in terms of how much text they contain, and how representative that text is of the actual content of the object. Note that the documents returned from a query do not have to be of the same type as the query object. For example, a component object can be relevant to a discipline object.

Comparing Documents using the Vector Model
Our starting point is to use the text vector model (Salton et al. 1995)[56]. For a collection of N documents and a total of n index terms (across the entire collection), a document matrix of size $N \times n$ is built. For each document-term element in this matrix, a weight w_{ij} is computed which represents the occurrence of term k_i in document d_j.

Essentially, each document is represented as a vector in the high-dimensional space of index terms. The similarity or relevance between two vectors can be quantified using the Euclidean distance, cosine, or dot product.

Three aspects can be taken into account when computing each weight w_{ij}: (1) the local frequency of term k_i in document d_j; (2) the global frequency of term k_i in the collection as a whole, and (3) the length of document d_j (so that short documents are not unreasonably favored or penalized).

---

[56] The three classic information retrieval models are the Boolean model, the vector model, and the probabilistic model. It is widely recognized that the Boolean model (which involves binary keyword matching) is too limiting because of its inability to recognize partial matches. There is ongoing discussion as to which of the two remaining models outperforms the other. For an overview and discussion of the classic information retrieval models, see Baeza-Yates and Ribeiro-Neto 1999, Chapter 2. CoMem adopts the vector model because of its relative simplicity and because of its useful extension, *latent semantic indexing*.

In CoMem, relevance is measured by calculating the cosine of the angle between any two document vectors. In other words, the documents are normalized by their length[57], and the relevance is always in the range [0,1]. It is important for the relevance to be bounded because it is mapped to a color in the visual displays of CoMem.

There remains however the question of which local and global term-weighting system to use. Experiments were conducted with three term-weighting systems:

- Binary (local: binary, global: none):
  $w_{ij} = 1$ if term k_i appears in document d_j
  $= 0$ otherwise

- Tf-idf (local: term frequency, global: inverse document frequency):
  $$w_{ij} = tf_{ij} \cdot \log \frac{N}{n_i}$$
  where tf_{ij} is the frequency of term k_i in document d_j, N is the total number of documents in the collection, and n_i is the number of documents in the collection in which term k_i appears.

- Log-entropy (local: log term frequency, global: entropy):
  $$w_{ij} = \log(tf+1) \cdot \sum_j \frac{\dfrac{tf_{ij}}{n_i} \log(\dfrac{tf_{ij}}{n_i})}{\log(N)}$$
  where tf_{ij} is the frequency of term k_i in document d_j, N is the total number of documents in the collection, and n_i is the number of documents in the collection in which term k_i appears.

Both tf-idf and log-entropy include global weighting factors that are intended to give less weight to terms that occur frequently or in many documents in the collection. The binary term-weighting system has no global component. Those three schemes were chosen out of the many possibilities and combinations because they cover the spectrum from simple to complex. In general, the performance of different weighting systems is highly dependent on the document collection and type of queries. It is very difficult to identify a scheme that consistently gives the best results[58].

Table 6 gives measurements of precision averaged over the 11 standard recall levels for each of the query sets and term-weighting schemes (with and without stemming).

---

[57] Calculating the cosine between two vectors is equivalent to normalizing the vectors so that they have unit length and then taking the dot product.

[58] The question of term-weighting systems is the focus of much research in information retrieval but is of only peripheral interest here. Salton and Buckley (1988) conducted experiments with various term-weighting systems for both queries and documents and found normalized tf-idf (i.e. cosine similarity between tf-idf vectors) to give the best average performance over 5 document collections, although its performance varied widely from collection to collection. Dumais (1991) conducted similar experiments but with latent semantic indexing (described below) rather than simple vector model analysis and found log-entropy to give the best results. Newer and more sophisticated term-weighting systems continue to appear in the literature (see for example Jung et al. 2000).

105

Table 6: Mean precision over the 11 standard recall levels using vector model comparisons for each of the query sets and term-weighting systems.

	No Stemming	Stemming
PROJECT QUERIES:		
Binary	0.31	0.31
Log-entropy	0.31	0.31
Tf-idf	0.31	0.31
DISCIPLINE QUERIES:		
Binary	0.39	0.45
Log-entropy	0.40	0.40
Tf-idf	0.37	0.41
COMPONENT QUERIES:		
Binary	0.49	0.49
Log-entropy	0.56	0.57
Tf-idf	0.60	0.63

Figure 53 gives precision versus recall curves for each of the query sets and term-weighting systems.

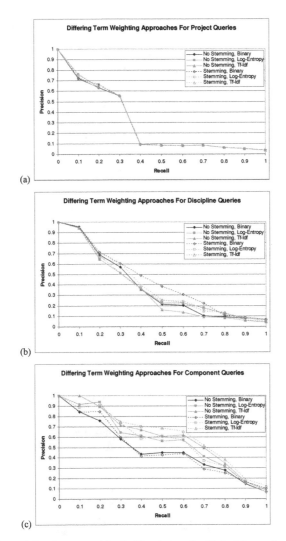

(a)

(b)

(c)

Figure 53: Recall versus precision for (a) project queries, (b) discipline queries, and (c) component queries using the vector model and various term-weighting systems.

The following observations can be made from the results of Figure 53:

- **Comparison of project, discipline, and component queries.** Overall, the vector model gives the best results for component queries and the worst results for project queries. This is not surprising since both project and component

107

documents usually contain only a few terms. However, in the case of component objects, these few terms are fairly representative of the content of the object. For example, "first floor (slab)" is a good indication of what that component is. On the other hand, the short annotations given to project objects are usually less representative of that project. For example, "Bay Saint Louis Hotel" says very little about that project. One way to improve performance in such cases is to consider the descendants of project objects when making comparisons involving those projects, an idea that will be explored later.

- **Term-weighting.** For discipline queries, binary weighting with stemming performs the best. This is because the other two term-weighting systems reduce the weights of the class terms since they occur frequently over the entire collection, even though the class terms give a better indication of the content of the discipline than the discipline name. For component queries (which will probably constitute the majority of queries in CoMem) tf-idf with stemming performs the best; however the differences are minor. When 90% confidence intervals are calculated for the mean precision achieved for component queries with each term-weighting system, these confidence intervals are found to overlap considerably[59].

- **Stemming.** For project queries, stemming makes little difference. For discipline queries however, stemming consistently gives a considerable improvement. This is because some discipline documents use singular nouns for the class list (beam, column, slab) whereas others use plural nouns (beams, columns, slabs). For component queries, stemming also improves the performance slightly, which is not surprising as a part of the component text is the class to which it belongs which is also sometimes plural and sometimes singular. The differences, however, are relatively slight. As above, 90% confidence intervals to compare weights with and without stemming overlap considerably.

Comparing Documents Using Latent Semantic Indexing

Latent semantic indexing (LSI[60]) is a refinement of the simple vector model that addresses the problems of synonymy (using different words for the same idea) and polysemy (using the same word for different ideas). LSI uses a technique called singular value decomposition to give a lower rank approximation of the original document matrix. The claim is that this approximation models the implicit higher order structure in the association between terms and concepts (Deerwester et al. 1990, Dumais 1991, Landauer and Dumais 1997). For example, if the two terms *beam* and *girder* frequently co-occur within documents in the collection or if they frequently occur in the same contexts, then an LSI query for *girder* would also return documents with only the term *beam*, an association that would be overlooked by the simple vector model.

---

[59] For example, stemmed binary weights give a mean precision of 0.49 with a 90% confidence interval of 0.30-0.67; stemmed tf-idf weights give a mean precision of 0.63 with a 90% confidence interval of 0.48-0.77.

[60] Also called *latent semantic analysis* in some contexts. Hill et al. (2001) use LSI for the related task of identifying shared understanding by analyzing design documents.

The input to LSI is an $N \times n$ matrix X of terms and documents as described above[61, 62]. This matrix X is decomposed into the product of three other matrices using singular value decomposition:

$$X = T_0 S_0 D_0'$$

T_0 and D_0 have orthogonal columns and S_0 is diagonal[63]. By convention, the diagonal elements of S_0 are constructed to be all positive and are ordered by decreasing magnitude. The approximating effect comes into play by keeping the largest k values from S_0 and setting the rest to zero. The product of the resulting matrices is a matrix $\acute{X}$ of rank k which is only approximately equal to X. A theorem due to Eckart and Young (in Golub and Reinsch 1971) suggests that $\acute{X}$ is the best rank-k approximation in the least squares sense to X. Since zeros were introduced into S_0 to obtain a new diagonal matrix, the representation can be simplified by deleting the zero rows and columns of S_0 to obtain a new matrix S, and then deleting the corresponding columns of T_0 and D_0 to obtain T and D respectively. The result is an approximation that, it is claimed, eliminates noise in the full model that impairs retrieval performance.

$$X \approx \acute{X} = TSD'$$

As with the full vector model, each column in $\acute{X}$ is a representation of a document, but in a space of reduced dimensionality. Documents can be compared by measuring the cosine between vectors as before. The performance of the reduced model is highly dependent on the amount of dimensionality reduction, i.e. the choice of k. This dimensionality reduction is henceforth referred to as the number of LSI factors and is examined in this chapter. Deerwester et al. (1990) and Dumais (1991) suggest a value of about 100, while Landauer and Dumais (1997), working with a different collection, found 300 to give the optimum performance. It is noteworthy that in both cases k is *much* smaller than the actual number of terms in the collection which is of the order of magnitude of thousands.

Several tests were conducted with LSI to determine whether it could offer an improvement over the performance obtained above with the simple vector model.

The data from the LSI runs is shown in Figure 54 (the "No Additional Corpus" lines). Each graph shows the mean precision over the 11 standard recall levels obtained by

---

[61] The description of LSI in the following paragraphs is based on Deerwester et al. 1990.

[62] The choice of term-weighting scheme used in this input matrix is an independent problem, although, as noted in the footnote above, log-entropy was found to give the best results (Dumais 1991). The results presented here were obtained using tf-idf weights as the input to LSI because these were found to give the best overall results in the simple vector model runs. Overall, it was found that the LSI results are not very sensitive to the type of weights used.

[63] As noted by Deerwester et al. (1990), singular value decomposition is closely related to the standard eigenvalue-eigenvector decomposition of a square matrix. T_0 is the matrix of eigenvectors of the square symmetric matrix $Y=XX'$ and D_0 is the matrix of eigenvectors of $Y=X'X$, and in both cases S_0^2 would be the matrix of eigenvalues. A version of LSI without singular value decomposition was developed by Wiemer-Hastings (1999) and found to give comparable performance to standard LSI.

running LSI using tf-idf weights and varying the number of dimensions. As before, separate results are presented for project, discipline, and component queries.

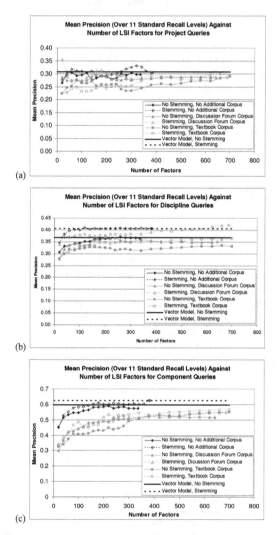

(a)

(b)

(c)

Figure 54: Mean precision over 11 standard recall levels for LSI runs against the number of LSI factors for (a) project queries, (b) discipline queries, and (c) component queries.

110

For discipline and component queries, LSI performs worse than the vector model for small numbers of factors, but gradually converges to the performance of the vector model as the number of factors is increased. For project queries, LSI gives a modest improvement over the simple vector model when 300-350 factors are used. As with the vector model runs, stemming consistently improves performance.

It is not surprising that LSI does not offer any significant improvement over the simple vector model. LSI is claimed to work best when the collection of documents is large and the documents are rich in keywords (in other words, the document matrix is dense). This helps the singular value decomposition, which is ultimately a statistical method, to infer relationships between terms based on their co-occurrence within individual documents more effectively. In this case, the document matrix is sparse. We have many documents but most of them consist of only two or three terms (for example the semantic label for a discipline object could be *Structure* or *HVAC*).

One way of addressing this problem is to add a set of "rich" documents alongside the CoMem documents. The rationale behind this is that if these additional documents are numerous enough and rich enough (i.e. contain many keywords), then LSI should be better able to infer relationships between terms because they frequently co-occur in the additional documents. These inferred relationships should in turn improve retrieval performance when comparing CoMem objects.

This approach was tested with two sets of additional documents:
- A collection of discussion forum messages exchanged by the design teams working on the projects in the experimental corporate memory. Each individual message was treated as a single document. Refer to the "Discussion Forum Corpus" lines in Figure 54.
- A set of articles from reference handbooks for professional structural designers and construction managers[64]. Each paragraph was treated as a single document. Refer to the "Textbook Corpus" lines in Figure 54.

It can be seen from Figure 54 that this did not improve the precision of LSI. In both cases, adding an additional corpus further reduced the performance of the LSI runs.

Context-Sensitive Comparisons: Concatenating Documents
As noted above, CoMem documents do not belong in a flat collection but are hierarchically structured. It would make sense therefore to consider an object's relatives within the hierarchy when making comparisons involving that object. Before

---

[64] These articles were: *Structural Analysis* by J. Y. Richard Liew, N. E. Shanmugam, and C.H. Yu (187 pages), *Structural Concrete Design* by Amy Grider, Julio A. Ramirez, Young Mook Yun (73 pages), *Structural Steel Design* by E. M. Lui (107 pages), *Construction Estimating* by James E. Rowings, Jr. (28 pages), and *Construction Planning and Scheduling* by Donn E. Hancher (31 pages). The handbooks in which these articles appear were available in electronic PDF form through the publisher, CRC Press, and Stanford Libraries.

developing a more sophisticated approach involving tree matching, the seemingly naïve approach of simply concatenating documents will be examined to see if it offers any improvement in retrieval performance. When converting an object to a document, an attempt will be made to include the text from that object's ancestors and/or descendants. Specifically, the following options are tried (with the rationale for each one described):

- **Concatenating descendants**: The retrieval performance for project queries is fairly weak. It has already been noted that the short label given to a project object is usually a poor indication of the content of the project. Concatenating the texts of all the project's descendants (disciplines and components) to the project document will result in a much longer text, which may improve retrieval performance. The same is also true of discipline objects. Although the problem of short, undescriptive labels is not so acute, concatenating the discipline's descendant components onto the discipline document might improve performance. A component objects does not have descendants and so the composition of the text will not change.
- **Concatenating ancestors**: A project object does not have ancestors and so the text of project documents will be unaffected. In the case of component and discipline objects, retrieval performance might be improved because objects belonging to similar parents (to the query item) will be ranked above those coming from unrelated parents. For example, if the query item is the *cooling tower* component from the *Bay Saint Louis hotel* project, then the *cooling tower* component from the *Las Vegas hotel* project will be ranked higher than the *cooling tower* component from the *Boston office building* project, because *Bay Saint Louis* and *Las Vegas* are both *hotel* projects.
- **Concatenating both descendants and ancestors**: Only discipline objects have both descendants (components) and ancestors (a project). This is included for completeness to determine whether it improves the retrieval performance compared to the other two options above.

Examples of concatenated documents are shown in Table 7 using the typical CoMem objects of Figure 52.

112

Table 7: Examples of concatenated documents using the objects of Figure 52.

Object Type	Object Text	Document Text – descendants concatenated	Document Text – ancestors concatenated	Document Text – both descendant and ancestors concatenated
Project	Project Name: Bay Saint Louis	bay saint louis structure slab column 1st floor slab b1 floor1 column the columns on column line b1 have been shifted to keep the classrooms column free	bay saint louis	bay saint louis structure slab column 1st floor slab b1 floor1 column the columns on column line b1 have been shifted to keep the classrooms column free
Discipline	Discipline Name: Structure Classes: {slab, column}	structure slab column 1st floor slab b1 floor1 column the columns on column line b1 have been shifted to keep the classrooms column free	bay saint louis structure slab column	bay saint louis structure slab column 1st floor slab b1 floor1 column the columns on column line b1 have been shifted to keep the classrooms column free
Component (without note)	Component Name: 1st floor Class: slab	1st floor slab	bay saint louis structure slab column 1st floor slab	bay saint louis structure slab column 1st floor slab
Component (with note)	Component Name: B1-floor1 Class: column Note: The columns on column line B1 have been shifted to keep the classrooms column-free.	b1 floor1 column the columns on column line b1 have been shifted to keep the classrooms column free	bay saint louis structure slab column b1 floor1 column the columns on column line b1 have been shifted to keep the classrooms column free	bay saint louis structure slab column b1 floor1 column the columns on column line b1 have been shifted to keep the classrooms column free

The results (mean precision over the 11 standard recall levels) from repeating the simple vector model analysis with concatenated documents are shown in Figure 55. Again, separate results are shown for the project, discipline, and component query sets. The "Simple" bars denote the performance without any concatenation and "Both" denotes the performance with concatenation of both ancestors and descendants.

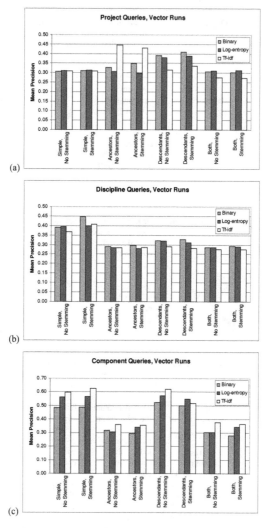

Figure 55: Mean precision over 11 standard recall levels for (a) project, (b) discipline, and (c) component queries using various forms of concatenation and term-weighting systems.

For project queries (Figure 55 (a)) concatenating descendants gives improved performance as expected. Concatenating descendants works best when stemming is combined with binary weights. The 90% confidence interval for stemmed binary

114

weights *without* concatenation of descendants is 0.29-0.33. The 90% confidence interval for stemmed binary weights *with* concatenation of descendants is 0.37-0.45. However, the concatenation of descendants is most effective for binary weights. It could be that the concatenation of descendants causes the global frequencies of important keywords to increase, in which case the "dampening" effect of the global components of log-entropy and tf-idf is undesirable. The best performance overall for project queries is for tf-idf weights applied to documents without stemming where the ancestors are concatenated (90% confidence interval 0.29-0.33 without concatenation, versus 0.40-0.49 with ancestors concatenated). This is because a significant proportion of the objects that should be returned by project queries (as judged by the human expert) are disciplines and components within the same project. Concatenating the project name to such disciplines and components enables them to be retrieved more effectively by project queries.

For discipline queries (Figure 55 (b)), any form of concatenation causes a decrease in performance. As noted earlier, the text of discipline objects (which consists of a name and a list of classes) already provides a very good representation of the content of the discipline. Additional terms from concatenated documents simply dilute the effectiveness of those representative discipline terms.

For component queries (Figure 55 (c)), concatenating ancestors and concatenating both descendants and ancestors both cause a decrease in retrieval performance. This may seem surprising, but can be explained by the fact that the ancestors of component objects (their disciplines and projects) were not given much consideration by the human experts when making human judgments of relevance. Interestingly, the concatenation of descendants causes a small improvement in overall performance. As noted earlier, component objects have no descendants and so component documents are unaffected by this concatenation, however the results will still differ because some of the returned results are disciplines or projects, which will be retrieved more effectively if their descendants are concatenated to them[65].

The LSI runs were repeated for concatenated documents in the hope that the corpus of concatenated documents would be richer and would enable associations to be detected based on term co-occurrence. As before, LSI does not offer any significant improvement over the simple vector model. Figure 56 shows a comparison of the project query performance between the vector model runs using tf-idf weights and the LSI runs using the optimum number of factors for each run. As before, LSI does not

---

[65] When comparing the performance of a component query with and without the concatenation of descendants, there are two possible sources of difference. The first is that discipline and project objects that should be returned by the query will have different compositions, even through the composition of the query component document itself will be unaffected by concatenation of descendants. The second is that the tf-idf or log-entropy weights in the document matrix might change due to increases in the global frequencies of some terms because of the concatenation process. The latter effect was found to be almost negligible. A similar observation can be made when comparing the performance of a project query with and without the concatenation of ancestors.

offer any significant improvements. Similar results were obtained for discipline and component queries.

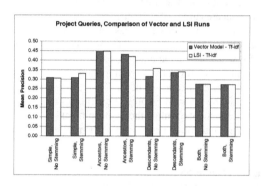

Figure 56: Mean precision over 11 standard recall levels for project queries. Comparison of the performance obtained with the vector model and that obtained using LSI with the optimal number of factors. Similar results were obtained for discipline and component queries, where the performance of LSI is very close to that of the vector model.

To summarize so far, this research experimented with the vector model using three different term weighting systems: binary, tf-idf, and log-entropy. Tf-idf weights performed the best overall. The tf-idf document matrix was used as the input to a series of LSI runs with varying amounts of dimensionality reduction. LSI did not offer any notable improvement over the simple vector model. Finally, an attempt was made to improve the performance of the simple vector model by concatenating related documents. In the case of project queries, using tf-idf weights and concatenating ancestors gave considerable improvements, as did using binary weights and concatenating descendants. Otherwise, concatenation did not help.

Context-Sensitive Comparisons: Tree Matching
At the beginning of the previous section, it was argued that it would make sense to consider an object's relatives in the hierarchy when making comparisons involving that object. The relatively simple method of concatenating the text from ancestors and descendants was tried. Here, a more elaborate method is developed, which is inspired by the concept of tree isomorphism[66].

The basic idea is that when comparing two objects, the comparison should try to find the best possible one-to-one match between those objects, their ancestors and their

---

[66] Two graphs G and G' are isomorphic if it is possible to label the vertices of G to be vertices of G', maintaining the corresponding edges in G and G'. In other words, there exists a one-to-one mapping between the vertices in G and those in G' such that if any two vertices are connected by an edge in G, then the mapped vertices in G' are also connected by an edge. *Tree isomorphism* is a special case involving trees rather than graphs.

descendants. The closeness of this match then becomes the relevance measure. This is slightly different than the classic isomorphism problem. When dealing with simple vertices connected by edges, the only factor which determines whether there is a match between vertex u in G and vertex v in G' is the topology of the two trees, i.e. the edges connecting those vertices. In this case, there is a notion of matching which is independent of relationships within the hierarchy: one CoMem object can closely match another if the two objects have very similar texts[67]. Furthermore, in this problem, two vertices can only be matched to one another if they occur at the same depth in the tree (i.e. a project object can only be mapped to another project object and so on).

The vector model measurements of relevance between any two nodes will be used as a *simple* measure of relevance and will be aggregated into *compound* measures of relevance that take account of the relatives (ancestors and descendants) in the tree. This is best explained by describing each type of comparison separately. There are six possible types of comparisons:

- Component-Component comparisons
- Discipline-Discipline comparisons
- Project-Project comparisons
- Project-Component comparisons
- Discipline-Component comparisons
- Project-Discipline comparisons

Component-Component Comparisons
The *simple relevance* between components c_i and c_j is taken as r_{c_i,c_j}, where r_{c_i,c_j} is the simple vector model relevance between them.

The *compound relevance* $\hat{r}_{c_i,c_j}$ between components c_i and c_j is:

$$\hat{r}_{c_i,c_j} = w_c r_{c_i,c_j} + w_d r_{d_i,d_j} + w_p r_{p_i,p_j}$$

where r_{d_i,d_j} is the simple relevance between c_i's parent discipline d_i and c_j's parent discipline d_j, and r_{p_i,p_j} is the simple relevance between c_i's grandparent project p_i and c_j's grandparent project p_j; and w_c, w_d, and w_p are weights such that $w_c + w_d + w_p = 1$.

---

[67] Pisupati et al. (1996) address a similar problem in which they match sub-trees based both on topology and on the location of the vertices in 3D space.

117

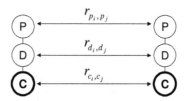

Figure 57: Compound component-component comparisons. The compound relevance is a weighted mean of the simple relevancies between the two components, the two disciplines, and the two projects.

In other words, the compound relevance between two components is a weighted mean of the component, discipline, and project relevancies (Figure 57). Note that all the simple relevancies are in the range $[0,1]$ and so the compound relevance will also be in the range $[0,1]$.

Discipline-Discipline Comparisons
The *simple relevance* between disciplines d_i and d_j is taken as r_{d_i,d_j}, where r_{d_i,d_j} is the simple vector model relevance between them.

The *compound relevance* $\hat{r}_{d_i,d_j}$ between disciplines d_i and d_j is:

$$\hat{r}_{d_i,d_j} = w_c g(d_i,d_j) + w_d r_{d_i,d_j} + w_p r_{p_i,p_j}$$

where r_{p_i,p_j} is the simple relevance between d_i's parent project p_i and d_j's parent project p_j, and $g(d_i,d_j)$ is some aggregated function of the simple relevancies between discipline d_i's m component children and discipline d_j's n component children. We can say without loss of generality that $m \leq n$.

The best way to think of $g(d_i,d_j)$ is as providing some aggregated measure of relevance between d_i and d_j based on simple relevancies between their children components. There are a total of $2mn$ possible directed edges between the set of d_i's m component children and the set of discipline d_j's n component children such that each edge spans the two sets. Each edge has a relevance value associated with it, which is the simple vector model relevance between the two components connected by the edge. We would like to find some subset of those edges which best represents the relevance between those two sets of components, and calculate the mean relevance associated with this subset. This becomes the value of $g(d_i,d_j)$.

For example, in the spirit of isomorphic tree matching, one option is to choose a subset of edges such that each component in the smaller set has one outgoing edge and each component in the larger set has no more than one incoming edge. In other words, the comparison will try to find the best one-to-one mapping from the components in the smaller set to the components in the larger set. There are P_m^n possible mappings (the

118

number of possible permutations of m taken from n). Each possible mapping can be represented by a set of m edges, and can be evaluated by taking the mean relevance of those edges. The value of $g(d_i, d_j)$ is the mean relevance of the best possible mapping. This is shown in Figure 58.

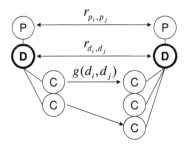

Figure 58: Compound discipline-discipline comparisons. The compound relevance is a weighted mean of the simple relevancies between the two projects, the two disciplines, and some aggregated function of the simple relevancies between the two sets of components. The most accurate method would be to find a one-to-one mapping between the two sets of components and then use the mean relevance of this mapping in the weighted mean.

This method is computationally demanding for large sets of components. Finding the best mapping means evaluating every possible mapping[68]. For m and n approximately equal, the number of possible mappings is of the order of $n!$. Larger CoMem disciplines can have more than 20 components; $20! = 2.4 \times 10^{18}$.

Therefore, for larger[69] sets of components an alternative method is used. In this case, a subset of edges is chosen such that the highest-relevance outgoing edge for each component in both sets is included. Subsets of this type will have $m+n$ edges, and can be evaluated by calculating the mean relevance of those edges, which is taken as the value of $g(d_i, d_j)$. This is shown in Figure 59.

---

[68] There may be intelligent ways of pruning this search space but they were not investigated.

[69] In the results presented below, a cutoff value of nine was used. If the larger of the two sets of components has more than nine components, the simpler method is used.

119

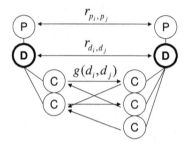

Figure 59: Simpler method for compound discipline-discipline comparisons. The compound relevance is a weighted mean of the simple relevancies between the two projects, the two disciplines, and some aggregated function of the simple relevancies between the two sets of components. In this simpler method, each component is mapped to the most relevant component from the other set. This is less accurate because it does not enforce a one-to-one mapping but is less computationally demanding.

The main advantage of this method is that it finds a local optimum with very little search. The main disadvantage is that it does not enforce a one-to-one mapping and so is arguably less accurate.

Project-Project Comparisons

The *simple relevance* between projects p_i and p_j is taken as r_{p_i,p_j}, where r_{p_i,p_j} is the simple vector model relevance between them.

The *compound relevance* $\hat{r}_{p_i,p_j}$ between projects p_i and p_j is:

$$\hat{r}_{p_i,p_j} = h(p_i,p_j) + w_p r_{p_i,p_j}$$

where $h(p_i,p_j)$ is some aggregated function of the simple relevancies between project p_i's discipline children and project p_j's discipline children which also takes account of the relevancies between the components in each discipline; and w_p is a weight such that $w_c + w_d + w_p = 1$ (w_c and w_d are applied inside $h(p_i,p_j)$).

If project p_i has m discipline children and project p_j has n discipline children, then (as before), there are a total of $2mn$ possible directed edges between the set of p_i's children and the set of p_j's children such that each edge spans the two sets. Each edge has a relevance value associated with it. This relevance value must take account of the simple relevance between the two disciplines as well as some aggregated relevance between their two sets of component children. For any edge between discipline d_i and discipline d_j, where d_i is a child of p_i and d_j is a child of p_j, the relevance value of that edge is:

$$w_c g(d_i,d_j) + w_d r_{d_i,d_j}$$

where $g(d_i,d_j)$ is an aggregated measure of the relevance between the two sets of component children as described above, r_{d_i,d_j} is the simple vector relevance between the two disciplines, and w_c and w_d, are weights such that $w_c + w_d + w_p = 1$.

120

As before, it is desired to find some subset of those edges which best represents the relevance between those two sets of disciplines, and calculate the mean relevance associated with this subset of edges. This becomes the value of $h(p_i, p_j)$.

As with the matching process between sets of components, a subset of edges is chosen such that each discipline in the smaller set has one outgoing edge and each discipline in the larger set has no more than one incoming edge. In other words, the comparison will try to find the best one-to-one mapping from the disciplines in the smaller set to the disciplines in the larger set (Figure 60). In the case of disciplines, there is no need to worry about the size of the search space because no CoMem project has more than six disciplines.

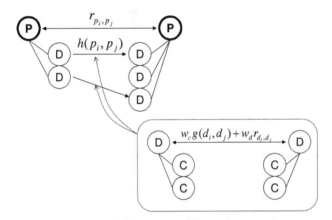

Figure 60: Compound project-project comparisons. The compound relevance is a weighted mean of the simple relevance between the two projects and some aggregated function of the relevancies between the two sets of disciplines. For each individual discipline-discipline comparison, a measure of relevance is used which is a weighted mean of the simple relevance between the two disciplines and some aggregated measure of relevancies between the two sets of components (lower right of figure).

Project-Component Comparisons

The compound relevance $\hat{r}_{p_i,c_j}$ between project p_i and component c_j is:

$$\hat{r}_{p_i,c_j} = k(p_i,c_j) + w_p r_{p_i,p_j}$$

where r_{p_i,p_j} is the simple relevance between project p_i and c_j's grandparent project p_j and $k(p_i,c_j)$ is some aggregated function representing the best possible mapping between project p_i's discipline children and component grandchildren and component c_j and its parent discipline d_j.

121

In this case, it is simpler to find the best mapping because the comparison is one-to-many rather than many-to-many (Figure 61). On the right hand side of Figure 61 there is component c_j and its parent discipline d_j. Those are compared to every possible component-discipline child-parent pair (d_i, c_i) on the left hand side of the figure where d_i and c_i are descendants of project p_i. The relevance of each match is evaluated using the following expression:

$$w_c r_{c_i,c_j} + w_d r_{d_i,d_j}$$

In other words, a weighted mean is evaluated of the simple vector model relevancies between the two disciplines and the two components being compared. The highest value of this weighted mean represents the best possible mapping and becomes the value of $k(p_i, c_j)$ which is used in the final compound relevance between p_i and c_j.

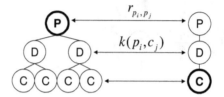

Figure 61: Compound project-component comparisons. The compound relevance is a weighted mean of the simple relevancies between the pairs of projects, between the discipline on the right and the set of disciplines on the left, and between the component on the right and the set of components on the left. In this case, the comparison finds the discipline-component parent-child pair on the left that best matches the discipline-component parent-child pair on the right. The comparison is one-to-many.

It is worth noting that the above formulation does not use the direct vector model similarity between the project object and the component object that are being compared. The two objects are of different types and therefore cannot be mapped to one another or directly compared. This is also true of discipline-component and project-discipline comparisons.

Discipline-Component Comparisons
The compound relevance $\hat{r}_{d_i,c_j}$ between discipline d_i and component c_j is:

$$\hat{r}_{d_i,c_j} = l(d_i, c_j) + w_d r_{d_i,d_j} + w_p r_{p_i,p_j}$$

where r_{p_i,p_j} is the simple relevance between d_i's parent project p_i and c_j's grandparent project p_j and r_{d_i,d_j} is the simple relevance between d_i and c_j's parent discipline d_j and $l(d_i, c_j)$ is some aggregated function of the simple relevancies between discipline d_i's children components and component c_j.

122

Again, it is simpler to find the best mapping because the comparison is one-to-many rather than many-to-many (Figure 62). On the right hand side of Figure 62 there is component c_j and on the left there is the set of the d_i's component children. The relevance between c_j and each child c_i is evaluated using the following expression:

$$w_c r_{c_i,c_j}$$

The highest value of this weighted mean represents the best possible mapping and becomes the value of $l(d_i,c_j)$ which is used in the final compound relevance between d_i and c_j.

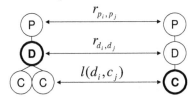

Figure 62: Compound discipline-component comparisons. The compound relevance is a weighted mean of the simple relevancies between the pairs of projects and disciplines, and between the component on the right and the set of components on the left. The component on the left that best matches the component on the right is found. The comparison is one-to-many.

Project-Discipline Comparisons
The compound relevance $\hat{r}_{p_i,d_j}$ between project p_i and discipline d_j is:

$$\hat{r}_{p_i,d_j} = m(p_i,d_j) + w_p r_{p_i,p_j}$$

where r_{p_i,p_j} is the simple relevance between project p_i and d_j's parent project p_j and $m(p_i,d_j)$ is some aggregated function of the simple relevancies between project p_i's discipline children and component grandchildren and discipline d_j and its component children.

The function $m(p_i,d_j)$ is similar to $h(p_i,p_j)$ used in the project-project comparisons. But whereas in the project-project comparisons two sets of disciplines were compared, here a single discipline is compared to a set of disciplines. Each possible mapping between d_j on the right hand side of Figure 63 and a child d_i of project p_i on the left hand side can be evaluated using the following expression:

$$w_c g(d_i,d_j) + w_d r_{d_i,d_j}$$

where $g(d_i,d_j)$ is an aggregated measure of the relevance between the two sets of component children as described above, r_{d_i,d_j} is the simple vector relevance between the two disciplines, and w_c and w_d, are weights such that $w_c+w_d+w_p=1$. The highest value of this expression becomes the value of $m(p_i,d_j)$.

123

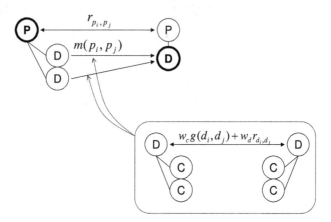

Figure 63: Compound project-discipline comparisons. The compound relevance is a weighted mean of the simple relevance between the two projects and some function of the relevancies between the discipline on the right and the set of disciplines on the left. The discipline on the left that best matches the discipline on the right is found. As before, for each individual discipline-discipline comparison, a measure of relevance is used that is a weighted mean of the simple relevance between the two disciplines and some aggregated measure of relevancies between the two sets of components (lower right of figure).

Tree Matching Retrieval Performance

How does the above tree matching formulation compare to the simple vector model when used as a measure of relevance in CoMem? Figure 64 shows the retrieval performance for both the simple vector model using stemming and tf-idf weights and the tree matching method where those same vector model comparisons are aggregated into compound relevance measures that take account of contextual objects.

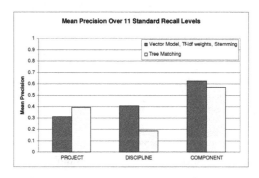

Figure 64: Mean precision over 11 standard recall levels for project, discipline, and component queries. Comparison of simple vector model and tree matching approach.

124

Tree matching outperforms the vector model for project queries (90% confidence interval 0.29-0.33 with vector model versus 0.36-0.43 with tree matching), but not for discipline or component queries. As noted earlier, project documents consist only of the project name, which is not highly indicative of the type of project, and so comparing a project to other objects based on that project's disciplines and components makes sense. However, the retrieval performance achieved using the tree matching method is comparable to that achieved by concatenating documents.

On the other hand, discipline documents include the discipline name (e.g. "*structural system*") as well as a list of classes (e.g. *beam, column, slab*). In this case, direct vector model comparisons using that discipline document are adequate and taking account of related nodes does not add value.

A similar argument can be made for component queries. The text of a component document is a good representation of that component and so there is not much need to compare contextual objects. The difference in performance between the vector model and tree matching is smaller for component queries than for discipline queries.

One disadvantage of the tree matching method is that when comparing two objects of two different types, those two objects are never directly compared to one another. For example, when comparing a discipline with the text "*slabs {pre-cast, post tensioned, composite}*" to a component with the text "*first floor slab*", those two texts are never directly compared using the vector model. Instead, the discipline's children are compared to the component, and the component's parent is compared to the discipline. To investigate the extent to which the poor performance of the tree matching method is due to this effect, the evaluation of the results was re-run with all such comparisons eliminated. In other words, only comparisons between objects of the same type are included in the evaluation. Figure 65 shows the retrieval performance of the two methods when this restriction is enforced.

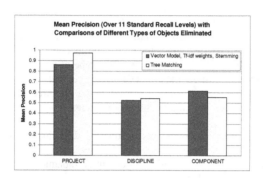

Figure 65: Mean precision over 11 standard recall levels for project, discipline, and component queries. Comparison of simple vector model and tree matching approach. All comparisons between different types of objects (e.g. project-discipline comparisons) are eliminated from the precision measurement.

In this case, the performance of the tree matching method is comparable to that of the vector model for all three types of queries. In particular, the tree matching method fares much better than before for discipline queries and outperforms the simple vector model for project queries. Overall, the performance of tree matching is comparable to the simple vector model but does not justify the extra computation it entails. Even for project queries, where tree matching performs considerably better than the simple vector model, a comparable improvement can be attained by concatenation which is much less computationally demanding.

Discussion and Closing Remarks

Reliability of the Results
Before drawing some conclusions from the above experiments, some comments are required about the reliability of the results presented here, particularly about their statistical significance. The experimental corporate memory with which these experiments were conducted contained 10 project objects, 35 discipline objects, and 1036 component objects. When evaluating the results, measures of mean precision were averaged for 8 project queries (i.e. with a project object as the query item), 18 discipline queries, and 6 component queries. In the case of project and discipline queries, the sample of queries tested represents a reasonable proportion of the population of all possible queries for my moderately sized corporate memory. For all three types of queries, the sample standard deviations were generally high, usually in the range 0.1-0.3. This is a reflection of the fact that the performance of information retrieval techniques can vary from query to query. As a result, it is difficult to make concrete conclusions about the performance of a system in general based on a sample of queries. In the case of project queries, 8 out of 10 possible queries were tested, and therefore the results for project queries carry much more weight. All the results were tested at the 90% significance level using the Student-t distribution, with the sample

126

standard deviation used as an estimate of the population standard deviation. In general, rather than making dichotomous decisions to reject or retain a null hypothesis, confidence intervals are quoted.

The second factor which may cast some doubt over the validity of the results is the fact that only one expert provided the "correct answer" for the sample queries, against which the results returned by the various measures were evaluated. In a more thorough study, two or more experts would provide correct answers and the level of agreement between the experts would be reported. This level of rigor was considered beyond the scope of this study.

Observations
Having noted these reservations, some tentative conclusions can be drawn. The most striking outcome is that there is no single relevance measure that consistently performs better than the rest. Furthermore, more complex relevance measures do not necessarily give better results than simpler ones.
- When comparing different term-weighting systems, the simplest system (binary weights) often had similar performance when compared to the most complicated (log-entropy).
- When taking context into account, the simple method of concatenating documents performed just as well as (if not better than) the more complicated tree matching method.
- When attempting to address the problem of synonyms, latent semantic indexing did not perform better than the simple vector model, whereas the relatively primitive dimensionality reduction achieved by stemming frequently performed better than both LSI and the unstemmed vector model.

Given these observations, what measures should be used in CoMem? The best overall performance is achieved using tf-idf weights in conjunction with stemming, and this is what is used by CoMem. For component and discipline queries, simple vector model comparisons are used. For project queries, the context (i.e. the discipline and component objects belonging to the project) needs to be taken into account, and concatenating the descendants of the project object onto the text of the project document is a simple and effective way of achieving this.

The tree matching approach as implemented here does not provide sufficiently improved performance to justify the additional computation it entails. It is worth developing further and refining in future research. In particular the choice of weights, w_p, w_d, and w_c, needs further investigation[70].

---

[70] Currently, the weights are chosen such that the levels being compared always receive a slightly higher weight. For example, when comparing two projects, w_p=0.5, w_d=0.25, and w_c=0.25; when comparing a project to a discipline, w_p=0.4, w_d=0.4, and w_c=0.2. A relatively simple trial-and-error analysis found this to give the best results. A more thorough approach would involve exploring the entire space of possibilities or using a machine learning method.

Why did LSI not Outperform the Simple Vector Model?

LSI did not deliver improvements over the simple vector model for the purposes of this research. The situations for which LSI has been shown to be effective are significantly different than ours. These differences are summarized in Table 8.

Table 8: Characteristics of corpora in which LSI was found to be effective, and the CoMem corpus, with and without additional documents added.

	MED standard collection [71]	CISI standard collection [72]	Groliers Academic American [73]	CoMem	CoMem plus Discussion Forum messages	CoMem plus technical reference articles[74]	CoMem – annotated objects only[75]
No. of documents	1033	1460	30,473	1081	1081+8125	1081+771	309
No. of unique terms in index [with stemming]	5823	5135	60,768	1541 [941]	16181 [9356]	5659 [5659]	1180 [856]
Mean no. of unique terms per document [with stemming]	50.1	45.4	151	5.7 [5.2]	26.8 [24.5]	20.8 [18.4]	13.3 [12.6]
Mean no. of unique terms per query [with stemming]	9.8	7.7	~1 (TOEFL synonyms test)	3.3 [3.2]	3.3 [3.2]	2.9 [2.8]	13.3 [12.6]
Mean no. of relevant documents per query	23.3	49.8	Not available	11.6	11.6	11.6	15.4

It can be seen from the table above that in the cases where LSI outperforms the simple vector model, the mean document size is significantly larger than in the CoMem corpus (refer to 'Mean no. of unique terms per document [with stemming]' row. It has been noted earlier that LSI makes statistical associations between synonyms because they repeatedly co-occur within the same document or they repeatedly occur in similar

---

[71] Deerwester et al. 1990, Dumais 1991

[72] Deerwester et al. 1990, Dumais 1991

[73] Landauer and Dumais 1997

[74] Textbook Corpus

[75] The subset of the CoMem corpus that was richly annotated, i.e. the items that had notes and change notifications linked to them.

contexts. In CoMem, objects are labeled with semantic annotations typically consisting of only a few terms, and therefore the document matrix is sparse rather than dense. The smaller the number of documents, and the more sparse the documents, the thinner the statistical sample from which LSI can make such associations. In the case of CoMem, the documents were simply too short, even with the addition to the corpus of discussion forum messages or technical articles[76].

To test the effect of document size, a smaller corpus was created consisting of only annotated CoMem objects. The statistics for this corpus are shown in the last column of Table 8. The mean document length for this smaller, richer corpus is 13.3 unique terms, which is significantly larger than the mean of 5.7 for the entire CoMem corpus (refer to 'Mean no. of unique terms per document [with stemming]' row). Again, discussion forum messages and technical articles were added to this smaller annotated corpus (the statistics for these are not shown in Table 8 for brevity). The results are shown in Figure 66. The overall performance is significantly improved (compare Figure 66 to Figure 54 (c)) but once again LSI fails to perform better than the simple vector model.

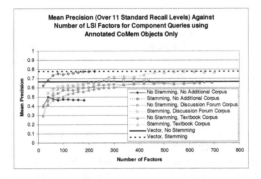

Figure 66: Mean precision over 11 standard recall levels for component queries in a reduced CoMem corpus where only richly annotated CoMem objects are included. The graph shows a comparison of the LSI performance at various numbers of factors versus the simple vector model performance.

To summarize, Figure 67 (a) below shows diagrammatically the behavior usually reported by advocates of LSI (Deerwester et al. 1990, Landauer and Dumais 1997). For some reduced number of dimensions, LSI outperforms the simple vector model. Figure 67 (b) shows diagrammatically the performance obtained with CoMem data using LSI. CoMem objects are shorter than regular documents used in typical LSI studies.

---

[76] Wiemer-Hastings (1999), using sentences as units of discourse (i.e. a single sentence is a document), also reports poor performance of LSI. Rehder et al. (1998), using LSI to grade student essays, report that if only the first 60 words or less of the student essay are used, then LSI performs poorly.

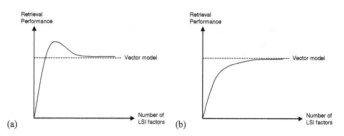

Figure 67: Diagrammatic representations of (a) the performance of LSI usually reported by its advocates, and (b) the actual performance of LSI obtained with the CoMem corpus.

Closing Remarks

The discussion of the relevance measures in this chapter has been based on the traditional assumption that given a user's information need and an information retrieval system, every retrievable item can be judged as "relevant" or "not relevant" and "retrieved" or "not retrieved". CoMem, with its emphasis on exploration rather than retrieval, is not a traditional information retrieval system and so this traditional assumption is not applicable, and consequently measures of precision and recall are not as important as they would be for more traditional systems. The relevance measures experimented with here will not be used to return a subset of the retrievable items in the form of a set of ranked "hits" to a query, but rather to highlight potentially relevant items in visual displays where most, if not all, of the corporate memory appears.

Finally, a comment may be made about the generalizability of the results. CoMem is hierarchical, as are most information systems: from the ubiquitous file systems of modern computers to more specialized information schemas in the construction industry such as IFC[77], AECXML[78], and so on. The short names given to CoMem objects are comparable to the names given to files and folders, or the names of objects in schemas such as IFC. The fact that relatively good retrieval results were achieved with such short texts (albeit not with LSI) is reassuring. Applying the vector model to such short texts and showing it to be effective is an important contribution of this research. Attempts to address this problem of short undescriptive texts by concatenating related documents or using tree matching are further contributions. As shown in Figure 66, the richer the texts, the better the performance, but even with the texts produced by basic state-of-practice systems available today, high retrieval performance is possible.

---

[77] See Caldas et al. (2002) for a similar research effort which attempts automatically to link construction documents to IFC components.

[78] See Lee et al. (2002) for a similar research effort based on XML documents.

Chapter 11

COMEM USABILITY EVALUATION[79]

Evaluation Approach
This chapter presents the evaluation of CoMem that assesses the extent to which it enables the designer to find and understand reusable items from the corporate memory, and the extent to which this ability to find and understand improves the effectiveness of the reuse process (Figure 9 in Chapter 4).

Since it is difficult to evaluate statements such as "designer can find and understand" or "external reuse is effective" in absolute terms, the strategy of this evaluation is to identify metrics for the validity of such statements and then to compare these metrics for CoMem versus "traditional tools", as shown in Figure 68. Traditional tools are tools that reflect the current state of practice of design reuse in industry. In addition, a set of variables are introduced into the comparisons to identify specific circumstances under which CoMem leads to more effective external reuse.

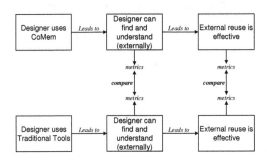

Figure 68: Approach for evaluation of CoMem. CoMem is compared to traditional tools in current practice.

CoMem Versus Traditional Tools
CoMem offers the following tools for *finding* and *understanding* items from the corporate memory:

- Overview (Corporate Map metaphor, Figure 69 (a))
- Project Context Explorer (fisheye lens metaphor, Figure 69 (b))
- Evolution History Explorer (storytelling metaphor, Figure 69 (c))

---

[79] Some of the contents of this chapter were first published in Demian, P. and Fruchter, R., 2006, "A Methodology for Usability Evaluation of Corporate Memory Design Reuse Systems." ASCE Journal of Computing in Civil Engineering, Volume 20, Issue 6, pp. 377-389. They are reproduced here with permission from ASCE.

131

The following tools were developed for the purpose of evaluating CoMem, and were used by the test participants as being representative of *traditional tools* used in current practice:

- **Outline Tree**. This is a prototype interface that uses indented lists of files and folders in the same way as Windows Explorer (Figure 70 (a)). The designer can use the Outline Tree to explore the corporate memory as if it were a set of files and folders on a computer, which reflects the nature of digital archives today, and the way current operating systems facilitate retrieval and exploration. It has an additional function to Windows Explorer: the generic icons for folders and files can be replaced by colored rectangles denoting the CoMem measure of relevance (the same relevance that is indicated on the CoMem Overview module, as shown in Figure 70 (b)). When the user selects an item from the Outline Tree, the versions of this item are displayed in a table similar to a spreadsheet or database program (Figure 70 (c)). The table displays the version number as well as the parent version and other ancestors. It can also display any textual information attached to that version (notes, notifications, and data).

- **Hit List**. This is a prototype web interface (Figure 71) that returns a list of hits in the same format as a web search engine, such as Google (Brin and Page 1998). Given a problem the designer is working on, he/she can bring up the Hit List at any time, and it will display a list of items from the corporate memory ranked by their relevance to the designer's current task (for exploration tasks). The user can also search the corporate memory by keyword, which is the mechanism expected to be used in the case of retrieval tasks. The user may select any item from the Hit List to display all the versions of that item in a web-based table similar to a spreadsheet or database program (Figure 71 (b)).

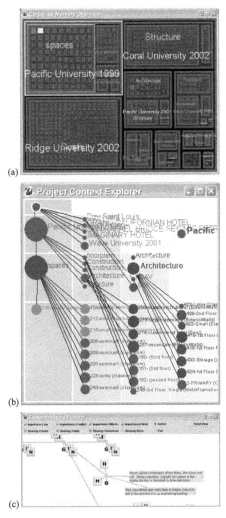

Figure 69: CoMem prototype. (a) CoMem Overview. (b) CoMem Project Context Explorer. (c) CoMem Evolution History Explorer.

133

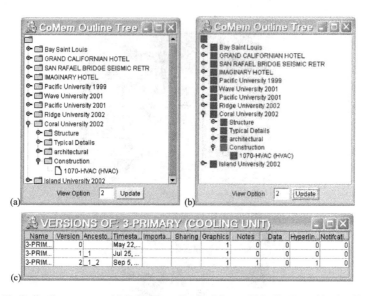

(a)

(b)

(c)

Figure 70: Outline Tree prototype. (a) Outline Tree with generic icons. (b) Outline Tree with colored icons to indicate relevance, used for exploration tasks. (c) Version table which lists all versions of an item in a table.

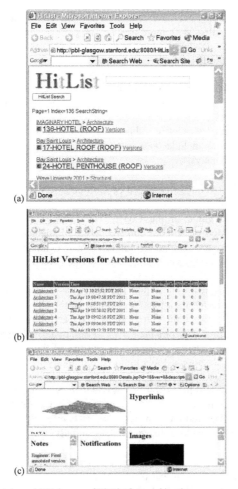

Figure 71: Hit List prototype. (a) Main page of Hit List for searching the corporate memory. (b) Web based version table. (c) Web view of an item select from Hit List.

Augmented Versus Unaugmented Traditional Tools

Some choices were made about what constitutes a set of traditional tools, such as:

- Does the user of "traditional tools" have access to evolution/version history data?
- Does the user of "traditional tools" have access to the automatically generated relevance measure, or must he/she rely on explicit queries such as keyword queries?

135

The advantage of including evolution and relevance data in the traditional tools is that it makes the evaluation of CoMem more precise: the comparison would be between two different *interfaces* for viewing the same data (i.e., only the interfaces are different, the underlying data is identical).

The disadvantage is that such data (history and relevance) is not commonly available in current practice – and so the "traditional tools" are not really representative of current practice. Ideally, the evaluation should measure the impact that evolution and relevance data would have on current practice – this is part of the value added by CoMem. One possible alternative is to include *both* unaugmented traditional tools *and* augmented traditional tools in the analysis.

- To measure the overall value added by CoMem, CoMem would be compared to unaugmented traditional tools.
- To measure the value of the CoMem *interface* (i.e. the modules and their metaphors), CoMem should be compared to the augmented traditional tools.
- To measure the value of having relevance and evolution data, the augmented traditional tools should be compared to the unaugmented traditional tools.

The hypothesis of this research is that the ability to find and understand leads to effective reuse. Augmented traditional tools improve the ability to find by providing a measure of relevance between the current design task and the contents of the corporate memory. CoMem further improves the ability to find by providing a succinct overview of the entire corporate memory which makes full use of the relevance measurements. Similarly, augmented traditional tools improve the ability to understand by providing evolution information. CoMem further improves the ability to understand by visualizing this evolution information in a single display as a coherent "story". There are therefore two components to an improvement in the "ability to find and understand": (1) the required data, in this case, relevance and evolution, is available, and (2) the user interface exploits this data in a meaningful way.

This evaluation focused on a comparison between augmented traditional tools and CoMem. It is taken as a given that providing relevance and evolution data improves finding and understanding. However, the more interesting aspect of "improving finding and understanding" which is investigated is the impact of CoMem's innovative interfaces that fully exploit relevance and evolution data.

Variables
The aim of this evaluation is not merely to determine whether CoMem offers improved support for reuse, but also to identify the specific circumstances under which traditional tools break down and CoMem offers genuine added value (and vice versa). The following variables are pertinent.

136

Type of finding task. There are two main kinds of finding tasks that need to be supported: *retrieval* and *exploration*.

- *Retrieval* occurs when the user is looking for a specific item: "I am looking for the cooling tower frame (component) from the structure (subsystem) of the Bay Saint Louis Hotel (project) that we worked on five years ago". Retrieval entails explicit queries by the user to express what he/she is looking for.
- *Exploration* occurs when the user has no idea what specific item to look for, only that it should be a relevant item related to the current design task or that it should satisfy certain conditions: "I am stuck trying to design a hotel cooling tower, is there anything in the system that can help me get started?" Exploration uses the user's design situation as an implicit query (Ye and Fischer 2002). In between the two extremes of retrieval and exploration there lie a whole range of tasks, for example when a user might have some notion that there is a specific item in the system that would be helpful, but cannot remember exactly where or what it is: "I remember designing a hotel cooling tower a few years ago... what project was that for and where in the system can I find it?"

Size of the repository. CoMem was designed with large repositories in mind, as this is where traditional tools often fail in supporting the finding and understanding of useful information. To what extent does CoMem also support smaller repositories, and what is the repository size for which traditional tools break down?

Metrics
The following metrics for *effective finding* were measured:

- Retrieval: Time to find the desired item. In the case that the user is looking for a specific item, the time taken to find that item is the most important metric.
- Exploration:
  1. Number of relevant items found. For each exploration task, an exhaustive list of useful items in the repository was prepared in advance by a human expert. This list was used to calculate a *recall score* for each test subject: the number of useful items found and listed by the user divided by the total number of useful items as judged by the human expert.
  2. The time taken to feel confident that the user has found everything there is to be found was measured. The test subject was instructed to continue exploring the corporate memory and listing all useful items until he/she felt that all useful items had been found.

The following metrics for *effective understanding* were measured:

- Ability to answer a set of questions after exploring the project context and evolution history, such as: "Why did the design team choose that building material?" A *context score* was generated for each user by dividing the number of correctly answered questions by the total number of questions asked. This was intended to measure the extent to which the tool enabled the user to understand

137

why that item was designed the way it was. The purpose of the questions was to test the ability of the user to *understand* content retrieved with that tool, rather than to test the user's domain expertise. Overly technical questions about architecture, engineering, or construction were avoided.

For *effective external reuse*, the extent to which the user agrees with the following statements was used as a measurable metric that assesses the effectiveness of the reuse process:
- "If I had this system in my work, I would reuse content from previous projects **more frequently** than I do currently."
- "If I had this system in my work, I would reuse content from previous projects **more appropriately** than I do currently."

CoMem was used in the context of synthetic experiments. If CoMem was used for a real project, possible metrics would have been:
- Percentage of designed artifact based on reused components.
- Quality of final design.

For each metric that was measured in the experiment, a 90% confidence interval was calculated and is displayed in the charts in this chapter. The Student-t distribution was applied, with the sample standard deviation used as an estimate of the population standard deviation and the number of degrees of freedom was estimated as the sample size minus one.

Method
Participants. Twenty participants were recruited from amongst students and researchers in the Department of Civil and Environmental Engineering at Stanford University, as well as professionals from local design offices. The participants were chosen to be as close as possible in age, computer experience, and design experience to eliminate any variability in the data due to these factors.

Materials. Three different software prototypes were tested:
- Outline Tree: indented list of projects, disciplines, and components, with versions of items displayed in tables.
- Hit List: web search engine with versions of items displayed in web-based tables.
- CoMem: Overview, Project Context Explorer, Evolution History Explorer.

Procedure
1. **Brief.** A standard passage describing each of the prototypes, the nature of tasks, and the objective of user tests was read to the participant (given in Appendix C).
2. **Warm up.** The participant was invited to familiarize himself/herself with the prototypes by exploring data unrelated to the tasks for about five minutes. During

138

this time, he/she was able to ask questions about how the prototypes work. After this warm-up, the formal experiment started.

3. **Retrieval tasks.** The participant was asked to complete three different randomly chosen retrieval tasks with CoMem, the Outline Tree, and the Hit List.

   Retrieval tasks are simple: "find the component called… which is in the discipline called… in the project called…". For each participant, the task selected to be completed using each prototype was randomly chosen. All retrieval tasks used were of comparable difficulty (for example, they were all component items from sub-trees of the corporate memory with similar branching factors). For each retrieval task the following were measured:
   - Time to complete the task, and
   - Correctness of final answer.

4. **Exploration tasks.** A standard passage describing a randomly-chosen synthetic scenario and a related exploration task based on the projects in the test bed repository was read to the participant. The participant was asked to explore the repository using CoMem and list all reusable items, until he/she feels confident that he/she has found all the reusable items in the repository. This was repeated for Outline Tree, and then the Hit List with different scenarios and tasks.

   Exploration tasks are of the type: "you are working on this problem, find anything you think would be helpful in the corporate memory to help you complete your design task." An example exploration task is shown in Figure 73, and the remaining five tasks are shown in Appendix B. There were a total of 6 previously-prepared exploration tasks, all of which were designed to be comparable in difficulty (for example, having the same number of reusable items and contextual questions). For each participant, the task chosen to be completed using each prototype was randomly chosen. The participant was asked to explore the corporate memory and make a list of all potentially reusable items found. After the task was completed, the participant was asked to answer some simple questions about each of the items listed, such as: "why did the design team choose that building material?". For each task, the following were measured:
   - **Recall score**: the proportion of potentially reusable items as judged by a human expert that were actually found by the participant.
   - **Context score**: the proportion of questions about helpful items that could be correctly answered by the participant.
   - **Time taken**: the time taken to feel confident that all helpful items had been found.

   The exploration tasks and the retrieval tasks were run first with a large repository, and then repeated with a small repository in the cases of CoMem and the Outline Tree.

5. **Questionnaire.** The participant was asked to complete three questionnaires, one for each of the prototypes, asking them about their subjective reactions to the prototype (shown in Figure 72, loosely based on Brooke 1996).

6. **Debrief.** Short, informal interview.

139

For both exploration and retrieval, the order of testing the three prototypes was randomly chosen, in an attempt to eliminate the effects of learning and increased familiarity with the data.

Questionnaire

The questionnaire given to test subjects to solicit subjective feedback on CoMem at the end of the test is shown in Figure 72. Test subjects were given similar questionnaires for Hit List and Outline Tree, but with questions 13, 14, and 15 omitted, as those questions are specific to CoMem.

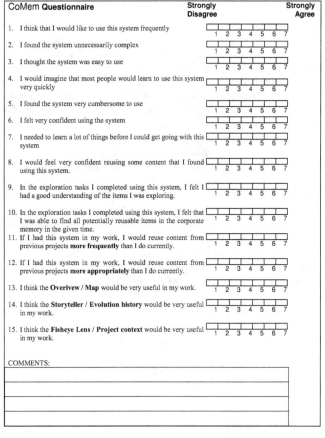

Figure 72: The CoMem questionnaire. Test subjects were given similar questionnaires for Hit List and Outline Tree, but with questions 13, 14, and 15 omitted.

The Data

The *large repository* tests described in this chapter were conducted on a pilot corporate memory consisting of 10 project objects, 35 discipline objects, and 1036 components. Of the 1036 component objects, approximately 30% were annotated with note objects. The *small repository* tests were conducted with the smallest possible subset of projects in the large repository that would include all the data required for the exploration tasks.

Attention was paid to ensure that the repositories were densely populated in several areas related to each exploration task. For example, if the exploration task involved roof design, care was taken to ensure that at least 5 or 6 projects had rich content related to roof design: annotations, hyperlinked documents, team interactions, images, design alternatives, and so on.

There was a pool of six standard exploration tasks from among which a task was randomly chosen for each prototype and repository size. Those were:

- Roof design
- Post-tensioned slab
- Shear walls
- Atrium
- Elevator
- HVAC System

Figure 73 below shows the first exploration task where the user is working on a roof design; the remaining five are shown in Appendix B .

TASK I = ROOF DESIGN ProblemIndex=136		
Reusable	**Reusable Items**	**Context Questions**
Pacific 1999	X	
Pacific 2001	Pacific 2001>Structure>449-Roof	What material was used for this roof? (Metal panels)
	Pacific 2001>Structure>444-PT Slabs	Why? (Lighter than concrete, simpler connections than steel, ease of construction)
	Pacific 2001>Construction>481-Roof System	What did the roof look like over the auditorium? (Pyramid) Why will the roof be expensive? (Because of the curvature) Which other building component had to be coordinated with the roof? (PT slabs)
Wave 2001	Wave2001>Arch>366-Roof	Can you name some architectural concepts that were considered? (Gable, mansard/French gable)
	Wave2001>Eng>363-Roof structure	What was the CM's feedback on the architect's ideas? (Complicatated, hieght restriction, snow/rain accumulating)
	Wave 2001>Construction>404-Air Handling Unit	What materials were considered? (Timber, steel)
		What equipment will go on the roof? (Air handling unit) What impact will this have on the structure? (Larger columns)
Coral 2002	Coral 2002>Structure>890-roof1(columns)	Can you describe the roof system? (prefab roof truss, elevated on columns and beams, prestressed roof slab)
	Coral 2002>Structure>888-roof1(beams)	Why was the roof truss elevated? (Natural ventilation, aesthetics)
	Coral 2002>Structure>892-slab1(roof) Coral 2002>Structure>894-rooftrusses1(rooftrusses)	
Ridge 2002	X	
Island 2002	X	
Bay Saint Louis	BSL>Arch>25-Ballroom (roof) BSL>Arch>17-Hotel roof (roof) BSL>Arch>24-Hotel Penthouse (roof)	
Grand Californian Hotel	GCH>Structure>59-Disney Store Roof (Steel dome) GCH>Structure>46-Area1 roof (roof truss) GCH>Structure>49-Area2 roof (roof truss) GCH>Structure>55-Area3 roof (roof truss)	
San Rafael Bridge Retrofit	X	
Imaginary Hotel	X	

Figure 73: A sample exploration task, where the user is searching for reusable items in roof design.

Retrieval Results

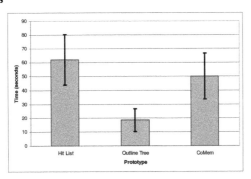

Figure 74: Time to complete a simple retrieval task with 90% confidence intervals displayed.

The time to complete a simple retrieval task is shown in Figure 74 above. The best performance in the case of retrieval was achieved by the Outline Tree which allowed retrieval tasks to be completed in the shortest time. The Outline Tree is effective for retrieval in the same way that binary search is effective for sorted arrays. By first selecting the project and discipline from much smaller lists than the list of all component objects in the corporate memory, the list of components that need to be visually scanned is greatly reduced. Further research should investigate the effectiveness of the Outline Tree for hierarchies with varying branching factors.

Pirolli et al. (2000) conducted closely related evaluations of visualizations of large tree structures. They did not include treemaps in their analysis, but compared Windows Explorer (equivalent to the Outline Tree) to hyperbolic trees. They conclude that the performance of the hyperbolic tree, because it attempts to crowd more data into a compressed space, is sensitive to "information scent" (the labels or colors used to guide the user to the appropriate piece of information).

After the Outline Tree, CoMem allowed retrieval tasks to be completed in the next shortest time. In spite of the fact that it was not developed with retrieval tasks in mind, CoMem still provides support for such tasks. Future research should investigate the role CoMem can play in retrieval tasks.

Exploration Results
The average time to complete an exploration task was comparable for the three prototypes CoMem, Outline Tree, and Hit List (14-18 minutes), even though, as discussed below, the user's performance in terms of recall score and context score varied considerably from tool to tool.

143

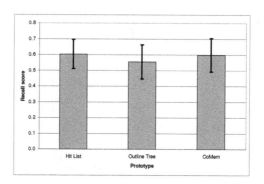

Figure 75: Recall score during exploration tasks with 90% confidence intervals displayed.

Figure 75 shows the fraction of relevant items successfully recalled by the test participants during exploration tasks (Baeza-Yates and Ribeiro-Neto 1999 gives a discussion of the measurement of *recall*). CoMem performed well in exploration recall. The Outline Tree had the poorest performance in exploration recall. This can be explained by the fact that in most cases reusable items were buried deep inside the hierarchy (i.e. at the component level) and left very little information scent at the higher levels that appear initially in the Outline Tree. Information scent is the user's perception of the value, cost, or access path of information sources. In the Outline Tree, projects and disciplines are displayed first and must be expanded by the user to display their component children. This requires that, for a relevant component, that component's parent discipline and grandparent project objects must also be relevant in order to encourage the user to expand those sub-trees and find the reusable component. This is rarely the case in the CoMem relevance measure.

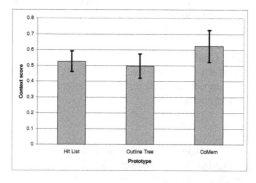

Figure 76: Context score during exploration tasks with 90% confidence intervals displayed.

Figure 76 shows the fraction of contextual questions that could be answered correctly by test participants about the items they retrieved. CoMem performed better than the Outline Tree and Hit List although it also had a slightly larger confidence interval. Most of the contextual questions were based on interactions between the designers, and the resulting version history of the item in question. The CoMem Evolution History Explorer was rated very highly by test participants. It was used during exploration tasks much more extensively than the Project Context Explorer, and was repeatedly praised by the participants during the debriefing interview.

Questionnaire Results

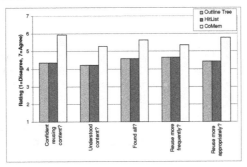

Figure 77: A selection of the questionnaire results.

Figure 77 shows the subjective opinions of the test participants about CoMem, the Outline Tree, and Hit List. For the questions regarding general usability characteristics (learnable, complicated, cumbersome), which are not displayed in Figure 77, CoMem attained comparable scores to the Hit List and Outline Tree. This is in spite of the fact that CoMem uses radically different interaction techniques, whereas the other two prototypes are tools with which any average computer user would be very familiar and experienced.

CoMem received higher scores particularly for questions 8-12 (Figure 72). Questions 11 and 12 are the main metrics for the extent to which external reuse is effective: does the user feel that if he/she had that prototype in his/her work, he/she would reuse designs more **frequently** and more **appropriately** (last two questions in Figure 77).

Questions 8, 9, and 10 (first three questions in Figure 77) measure the user's perceived ability to find and understand:
- "I would feel very confident reusing some content that I found using this system."
- "I had a good understanding of the items I was exploring."

145

- "I felt that I was able to find all potentially reusable items in the corporate memory in the given time."

The high score awarded to CoMem in these questions supports the higher recall and understanding performance measures achieved by the test subjects when using CoMem for exploration tasks.

The users were asked to rate the three CoMem modules: the Overview, the Project Context Explorer, and the Evolution History Explorer. The highest-rated module is the Overview, which validates the claim that providing a succinct overview of the entire corporate memory is extremely valuable, and that a treemap is a good visualization for this purpose. The Evolution History Explorer was also rated very highly. By observing the users during the tests, it is clear that this module enables the users to reconstruct the evolution of the designs and understand the rationale behind this evolution much more effectively than a list of versions or displays of single versions one at a time. The lowest-rated module, although by very slightly, is the Project Context Explorer. Many users found it unclear because it shows the same items as those in the Overview, but positioned and colored differently. Further development is needed to couple the Project Context Explorer more tightly with the Overview, so that a change in one display triggers a corresponding change in the other. It is suspected that advanced users of CoMem would make more use of the Project Context Explorer.

Discussion
At a global (macro) level, the results test the hypothesis of this research. Traditional tools do not support the ability to *find* and *understand* and traditional tools do not lead to effective reuse. CoMem supports the ability to *find* and *understand* and CoMem leads to effective reuse. This supports the claim that the steps of *find* and *understand* lead to effective reuse, as shown in Figure 78.

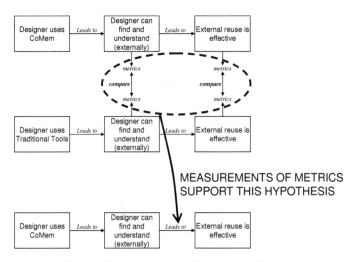

Figure 78: Macro evaluation to test the hypothesis of this research.

At a micro level, a comparison between the metrics from CoMem and those from traditional tools helps to identify the specific circumstances under which CoMem performs better than traditional tools. The first variable in this evaluation is the type of task: exploration versus retrieval. CoMem performs best in exploration scenarios.

The other variable that was introduced into the evaluation is repository size.

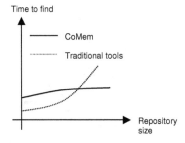

Figure 79: A diagrammatic representation of the hypothesized effect of repository size on the performance of information tools.

Figure 79 shows the hypothesized effect of repository size on the performance of CoMem and traditional tools. Figure 80 and Figure 81 show the actual effects observed on exploration time and retrieval time. In the case of exploration (Figure 80), the size of the repository seems to have little effect. A more subtle aspect such as the amount of

147

text that needs to be read to complete the task is more likely to have an effect on exploration time than the relatively simple count of the number of items in the repository. In the case of retrieval (Figure 81) the results are more similar to the hypothesized effect. As the repository size is increased, the performance of CoMem is assumed to stay approximately constant[80], while that of the Outline Tree begins to deteriorate (takes more time for the larger repository). By simple extrapolation, it can be imagined that a point would be reached beyond which CoMem outperforms the Outline Tree.

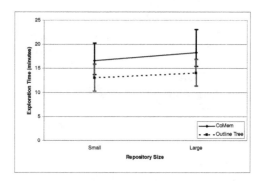

Figure 80: The effect of repository size on exploration time with 90% confidence intervals displayed.

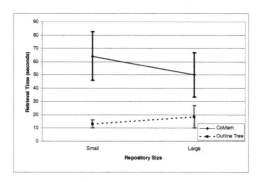

Figure 81: The effect of repository size on retrieval time with 90% confidence intervals displayed.

---

[80] As can be seen from Figure 81, the retrieval time is actually shorter for the larger repository. It can be seen from the 90% confidence interval that the reliability of this result is uncertain and logic dictates that it would take just as long, if not longer, to retrieve an item from a larger repository as from a smaller repository so it can be assumed that the performance of CoMem is approximately constant for both repository sizes.

148

Chapter 12

DISCUSSION AND CONCLUSIONS

This chapter presents a final discussion of this research as a whole, and particularly the results in light of the stated research hypothesis and research questions. It highlights the contributions of this research, and discusses the conclusions that can be drawn from it.

Discussion of Results

The objective of this research is to *improve* and *support* the process of design knowledge reuse in the AEC industry. Ethnographic observations show that the three key activities in *internal* knowledge reuse process are:

- *Finding* a reusable item
- Exploring this item's *project context* which leads to *understanding*
- Exploring this item's *evolution history* which leads to *understanding*

The hypothesis is that if the designer's interaction with the external repository enables him/her to:

- Rapidly *find* relevant items of design knowledge
- View each item *in context* in order to *understand* its appropriateness, specifically:
  - Explore its project context
  - Explore its evolution history
- ⇨ Then the process of reuse will be improved.

This improved reuse will lead to higher quality design solutions, and save time and money.

Based on this hypothesis, this research addressed the following questions:

Question 1: How does *finding* occur in internal knowledge reuse? What retrieval mechanisms are needed to support the *finding* of reusable design knowledge in a large corporate repository of design content? What are suitable interaction metaphors and visualization techniques?

Question 2: What is the nature of the project context exploration in internal knowledge reuse? How can this exploration be supported in a large corporate repository of design content? What are suitable interaction metaphors and visualization techniques?

Question 3: What is the nature of the evolution history exploration in internal knowledge reuse? How can this exploration be supported in a large corporate repository of design content? What are suitable interaction metaphors and visualization techniques?

The internal knowledge reuse aspects of these questions were addressed through the ethnographic study presented in Chapter 3. Internal knowledge reuse can be formalized into *finding* and *understanding*. Finding occurs by simultaneously comparing data at the three levels of granularity: project, discipline, and component.

The CoMem Overview (Chapter 6) explores how finding reusable design knowledge may be supported in external repositories using an innovative graphical user interface. The Corporate Map presents a succinct snapshot of the entire corporate memory that enables the user to make such multi-granularity comparisons and quickly find reusable items. In order to provide direct value to the users, and their search tasks, items on the map are color-coded based on the relevance analysis results. Chapter 10 presents an in-depth study of how this relevance may be measured.

The *understanding* step in internal reuse occurs through the six degrees of exploration described in Chapter 3. Chapter 7 and Chapter 8 address how this exploration can be supported in external knowledge reuse using a *Project Context Explorer* and *Evolution History Explorer*, as well as interaction metaphors and mechanisms for those modules. The Evolution History Explorer draws from the effectiveness of comic books for telling stories, and explores how this effectiveness can be carried over to the presentation of version histories. The Project Context Explorer combines the relevance measure with the classic fisheye formulation to aid the user in identifying and exploring related items in the corporate memory.

The usability evaluation results presented in Chapter 11 support the hypothesis of this research, that the ability to find and understand does lead to more effective reuse. CoMem offers greater support for finding and understanding than traditional tools, and reuse using CoMem is consistently rated to be more effective by test participants.

Contributions
The main contribution of this research is the recognition that reuse consists of the two tasks of *finding* and *understanding*, and the formalization of the reuse process. This formalization not only allows the development of an external reuse system, but also enhances internal reuse. An ensuing contribution is the *decoupling* of find and understand, in terms of the tasks that need to be supported, interaction metaphors for supporting these tasks, and processing of the knowledge in the corporate memory to facilitate finding and understanding.

The CoMem prototype constitutes a substantial contribution to information technology in the form of an innovative design of human-computer interface. The domains that it can be applied to are not limited to engineering design, but CoMem can be generalized to the task of finding and using content from large hierarchical repositories.

The CoMem relevance measure amounts to a significant contribution in the field of information retrieval. This research shows that even with the sparse and short-text data that occurs in real-world domains, adequate precision and recall performance is possible. In particular, CoMem makes the most of hierarchical relationships in the corporate memory. The tree matching approach inspired by tree isomorphism is an innovation in the field of information retrieval. This contribution is further amplified by CoMem's visual representation of relevance in cutting-edge interaction designs: CoMem puts the relevance measure to full use and supports exploration rather than retrieval.

The spectrum between exploration and retrieval is underlined in this research. Retrieval is disproportionately favored and exploration is commonly neglected in traditional tools. CoMem attempts to rectify this imbalance, by recognizing the importance of exploration, and appreciating the radically different interfaces that are needed to support it.

This research also makes a methodological contribution through the evaluation of CoMem. The CoMem usability evaluation represents a useful framework for evaluating information interfaces. The same data can be explored using different interfaces. Hit List, Outline Tree, and CoMem cover the spectrum of information interfaces, from traditional to innovative. The important dimensions of the evaluation space are the size of the repository, the type of task, and the user's familiarity with the data. Search engines and expandable/collapsible folder trees can be used to represent traditional information interfaces.

Conclusions

CoMem started with the observation that, whereas designers reusing designs from their personal experiences (*internal memories*) is an extremely effective process, designers reusing designs from digital or paper archives of content from previous projects often fails. From extensive ethnographic studies of practicing designers, this research identifies two reasons for the effectiveness of internal knowledge reuse:

1. Even though the designer's internal memory is usually very large, he/she is always able to *find* relevant designs or experiences to reuse.
2. For each specific design or part of a design he/she is reusing, he/she is able to retrieve a lot of contextual knowledge. This helps him/her to *understand* this design and apply it to the situation at hand. When describing contextual knowledge to others, the designer explores two contextual dimensions: the *project context* and the *evolution history*.

Armed with these observations, CoMem was developed to serve as an external reuse system that would enable designers to:

1. *Find* reusable items in large corporate archives
2. Explore the project context of these items in order to *understand* them
3. Explore the evolution history of these items in order to *understand* them

151

Based on the three reuse steps identified above – find, explore project context, explore evolution history – CoMem has three corresponding modules: an Overview, a Project Context Explorer, and an Evolution History Explorer.

Future Research

From anecdotal evidence observed during the user tests, the labeling of treemaps plays a very important role in their support for retrieval tasks. Very few of the test subjects used the keyword search function in CoMem during the retrieval tasks. Further research is needed to develop the labeling of treemaps and to understand the role of labeling in retrieval.

Further work is also needed to develop the relevance measure. This research paves the way for exciting innovations in information retrieval from large hierarchical information sources.

CoMem is poised to be generalized to a wide variety of domains. Work is already underway on an interactive workspaces version of CoMem which runs in technology-rich spaces with computing and interaction devices on many different scales (Johanson et al. 2002, Fruchter et al. 2007). CoMem prototypes are being developed for search in textual databases. New functions are being added that exploit concepts from the merging field of chance discovery. This research has laid the foundation for stimulating future research into knowledge capture and reuse, treemaps, measuring relevance, and evaluating information interfaces.

More work is required to investigate the effect of familiarity with the contents of the repository. CoMem must support novice users who are unfamiliar with the contents of the corporate memory as well as advanced users who are able to formulate explicit queries. In practice it will be impossible to be completely familiar with the corporate memory because it is constantly growing and evolving. Further studies should focus specifically on the user's familiarity.

It is hypothesized that traditional tools rely on the user's familiarity with the data to formulate explicit queries. CoMem should be less sensitive to familiarity and therefore provide greater support for novice users (Figure 82). The effect of familiarity can be studied by conducting two rounds of testing. Test participants would be chosen who are unfamiliar with the data. They would be asked to complete one set of tasks, then given a "familiarity-building exercise", and then asked to complete a second round of tasks. Their performance in the first and second sets of tasks would be compared to investigate whether the effectiveness of the tool is dependent on familiarity with the data.

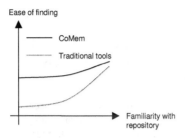

Figure 82: A diagrammatic representation of the hypothesized effect of familiarity with the data on the performance of information tools.

The three dimensions of size, task, and familiarity together define a three dimensional space (Figure 83). It is suspected that CoMem is particularly supportive in one corner of this space, while traditional tools support the opposite corner.

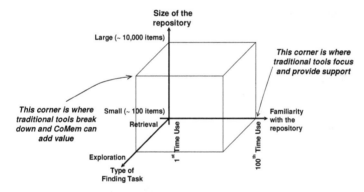

Figure 83: Three dimensional space defined by size, task, and familiarity. CoMem and traditional tools support opposite corners of this space.

For further readings on this research, refer to Demian and Fruchter 2005, 2006a, 2006b, 2006c, and Fruchter and Demian 2005, 2002, 2002(a), 2002(b), 2002(c).

153

EXAMPLE OUTPUT OF SCENARIO-BASED DESIGN METHODOLOGY

This appendix presents samples of the observations, scenarios, and analysis used during the scenario-based design of CoMem. Only one sample scenario is presented here. There were two others as described in Chapter 4, page 43. There were also multiple versions of each scenario and refinements based on the claims analysis.

Table 9: Guide for interviewing practitioners during the ethnographic study.

Guide for Interviewing Designers in the Design Office

Goal: To understand how designers reuse knowledge from past projects when working on a current project, from:

- Internal memory
- External memory (digital files, paper drawings, paper binders, etc.)
- Mentors or colleagues

Open-ended questions about reuse

Where would you say knowledge resides in your company?

Where do you go if you have a question?

How important do you think it is to capture and reuse knowledge? Are you willing to do extra work to capture knowledge?

Specific things about reuse

Guide for Interviewing Designers in the Design Office

Frequency and type of reuse

How frequently do you consciously reuse designed components or subcomponents from previous projects: From memory? From external archive? By asking mentor?

How frequently do you consciously reuse rules of thumb or domain expertise acquired from previous projects: From memory? From external archive? By asking mentor?

How frequently do you consciously reuse design tools (e.g. spreadsheets) acquired from previous projects: From memory? From external archive? By asking mentor?

External archives

Where is knowledge from previous projects stored?

Do you ever refer back to these archives?

What are the procedures for retrieving stuff from these archives?

Mentors

Is there anyone you work with whom you would consider a mentor?

Would you rather ask a knowledgeable colleague or mentor than look in the archive?

Reuse satisfaction

How satisfied are you with your experiences getting knowledge from: The archive? A knowledgeable colleague or mentor?

Table 10: Stakeholder profiles developed from the field study.

Stakeholder	General group characteristics
Novice designers (net knowledge consumers)	Background: Little or no design experience. Usually young age, 20s or 30s. Familiar with IT, web, e-mail, chat, forums. Expectations: Ability to find information/knowledge quickly. Complete, rich information/knowledge. Digital rather than analog/physical media. Reuse as a learning resource, rather than a productivity tool. Preferences: Comfortable with PC Windows, AutoCAD, Microsoft Office. Knowledge reuse should be fast and simple, secondary to design.
Experienced designers (net knowledge producers)	Background: Years of design experience. Usually older, 40s or 50s. Limited familiarity with IT. Expectations: Ability to do work without worrying about knowledge capture overhead (experienced designer will not rely on reuse system as much). Reuse as productivity tool (particularly CAD) rather than learning resource. Preferences: AutoCAD, paper drawings and documents.

Stakeholder	General group characteristics
Mentors, managers	Background: Many years of design experience. May be founder of design office. Delegates and coordinates work among other designers in the practice. Expectations: Reuse system should retain knowledge in the design practice. Process should not be completely automatic. Mentor still has a role in advising novices on what and how to reuse. Contextual knowledge to help novices learn from archives. Security and persistence of knowledge (not accidentally overwritten or deleted). Some control of storage and organization of knowledge. Preferences: Familiar technology, e.g. FTP site.

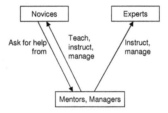

Figure 84: Relationships among stakeholders.

Table 11: Summary of themes from the ethnographic study.

Workplace theme	Issues contributing to the theme
Finding paper drawings	Drawings not easily accessible.
	Once found, contain wealth of information.
	Can find unexpected gold nuggets, serendipitously, while browsing.
	Once the project is identified (i.e. the set of drawings), it becomes possible to effectively find what you are looking for by following "hyperlinks".
Finding electronic drawings (and other files)	Typically using Windows Explorer, good for navigation, assuming you have a rough idea of what you are looking for.
	Large number of files per folder.
	Xrefs complicate things.
	Filenames not descriptive (tradeoff descriptive versus short, sortable, simple).
Making sense of drawings	Must understand visual vocabulary.
	If reuser was not involved in project, difficult to remember.
	Rationale not captured on drawing (mentor interviewed would argue that it is – through, for example, load key sheets).
Preference for reusing own designs	Designers always prefer to reuse their own designs.
	Sets a limit on what is reusable.
	Lack of context to blame?

Table 12: Hypothetical stakeholders used in the scenarios.

Hypothetical Stakeholders
Matthew (mentor) is the founder and CEO of "X Inc". He started the company in 1990 after completing his PhD at Stanford. Before coming to Stanford he had had over ten years of structural design experience. He is committed to the profession of structural engineering, its ethics and professional code of conduct. He is also passionate about training and mentoring the young engineers at his company.
Eleanor (expert) is an experienced structural engineer. She joined "X Inc" when it was founded ten years ago. She had just graduated from the Master's program at Stanford.
Nick (novice) is a novice engineer. He graduated a few months ago from the University of Colorado, and has been at "X Inc" since then.

Table 13: Sample problem scenario.

Problem Scenario
A mentor structural designer, Matthew, and a novice, Nick, both work for a structural design office in Northern California. The office is part of the "X Inc" Structural Engineering Firm. They are working on a ten-storey hotel

that has a large cooling tower unit. Nick must design the frame that will support this cooling tower. Nick gets stuck and asks Matthew for advice. Matthew recalls several other hotel projects that were designed by "X Inc". He lists those to Nick and tells him that the Bay Saint Louis project, in particular, would be useful to look at.

Matthew walks with Nick to the room where old paper drawings are kept. Together they locate the set of drawings for the Bay Saint Louis project. Matthew takes out the structural drawings and briefly explains the structural system of the building to Nick. Matthew then finds the specific drawing sheet with the Bay Saint Louis cooling tower frame detail.

The drawing shows the cooling tower frame as it was finally built. It is a steel frame. Matthew realizes that what he had in mind for Nick to reuse is an earlier version that had a steel part and a concrete part. He is not sure if this earlier version is documented somewhere in the archive. Rather than go through the paper archive again, Matthew simply sketches the design for Nick. Matthew's sketch also shows the load path concept much more clearly than the CAD drawing would have, which helps Nick to understand the design. Matthew explains to Nick how and why the design evolved. Given the current project they are working on, it would be more appropriate to reuse the earlier composite version. Matthew recalls that the specifications of the cooling tower unit itself, which were provided by the HVAC (heating, ventilation and air conditioning) subcontractor, had a large impact on the design. Nick now feels confident enough to design the new cooling tower frame by reusing the same concepts as the Bay Saint Louis cooling tower frame, as well as some of the standard details.

Table 14: Claims analysis of the problem scenario.

Situation posture	Possible pros or cons of the feature
Paper archives	+ Practitioners in current practice are used to this.
	+ (CAD) paper "hyperlink" trail seems to be very effective (starting from the general notes sheets one can follow 3-4 "hyperlinks" and get to any detail).
	− Difficult to search (e.g. keywords).
	− Expensive to produce, copy, transport.
	− Difficult to capture, store, index (automatically).
	− Sometimes lacking contextual information (although this is a more serious problem with digital archives − CAD drawings often have markups or scribbles on them.
Digital archives	+ Easy to copy, modify, transport (e.g. CAD).
	+ Easier to search, using computer algorithms, indexes, etc.
	− Can lack informal knowledge (difficult to capture).
	− Can require additional effort to capture and organize drawings and other content electronically.
Novice-mentor relationship	+ Novices can get personalized help when they need it.
	+ The mentor offers individual care.
	+ The mentor provides contextual information which may not be captured on the drawing, and answers questions.
	+ The mentor helps the novice to find reusable content, but also conveys the expertise which produced that content (by answering questions, etc.).
	+ The mentor knows what is reusable and where to look for it.
	+ Mentoring is a rewarding experience for the mentor (should not completely automate).
	− Novices in trouble if experienced designers unavailable.
	− Mentor may not have time.
	− Mentor may experience memory loss, or may not be familiar with the entire corporate memory.

Table 15: Metaphors for objects and activities in a corporate memory system for reuse (based on feedback from industry practitioners).

Activity	Real world metaphor	Implications for CoMem activities
Looking for a drawing in the paper archive is like…	Finding a needle in a haystack	Difficult to find reusable items in large archives

160

Activity	Real world metaphor	Implications for CoMem activities
Looking for a drawing (CAD file) in the hard disk is like…	Windows Explorer, depth-first search, breadth-first search, searching the web using a search engine	Must allow for different search strategies: exploring, submitting queries, browsing
Reading the notes on a drawing is like…	Eavesdropping	Provide anecdotal information
Browsing through versions of a drawing is like…	A flipbook	Quick browsing of versions
Being mentored by an expert engineer is like…	Learning by example, being spoon-fed, being led by the hand	Provide anecdotal information about design evolution, and experiences from previous projects
Mentoring a novice engineer is like…	Telling a story	
Going from a plan or section to a detailed drawing is like…	Navigating hypertext using hyperlinks, zooming in CAD	Provide tools for zooming in and out on bigger or smaller grains of design
Looking at a detail and recalling its project context is like…	Remembering the interactions (arguments) between the project team members	Describe interactions between project team members

Table 16: The problem scenario transformed into an activity scenario.

Problem Scenario	Activity Design Scenario
<Background on Nick, his motivations, …>	<Background on Nick, his motivations, …>
Nick gets stuck and asks Matthew for advice. Matthew recalls several other hotel projects that were designed by "X Inc". He lists those to Nick and tells him that the Bay Saint Louis project, in particular, would be useful to look at.	Nick gets stuck, but Matthew is not around to help. Nick decides to use the company's corporate memory system, CoMem. He identifies his current project to the system and CoMem lists some similar projects for him to look at.
Matthew walks with Nick to the room where old paper drawings are kept. Together they locate the set of drawings for the Bay Saint Louis project. Matthew takes out the structural drawings and briefly explains the structural system of the building to Nick. Matthew then finds the specific drawing sheet with the Bay Saint Louis cooling tower frame detail.	Nick chooses to explore the Bay Saint Louis project. He "zooms in" on the structural system for that project. With the structure in front of him, he zooms in further on the cooling tower frame.
The drawing shows the cooling tower frame as it was finally built. It is a steel frame. Matthew realizes that what he had in mind for Nick to reuse is an earlier version that had a steel part and a concrete part. He is not sure if this earlier version is documented somewhere in the archive. Rather than go through the paper archive again, Matthew simply sketches the design for Nick. Matthew's sketch also shows the load path concept much more clearly than the CAD drawing would have, which helps Nick to understand the design. Matthew explains to Nick how and why the design evolved. Given the current project they are working on, it would be more appropriate to reuse the earlier composite version.	Nick sees the evolution of this cooling tower frame as a series of versions. He can also see the notes and messages that were exchanged between the team members in generating these versions. He sees that the final version of the frame was steel. His current project has a concrete structure. Nick flips through the earlier versions and notices that some of them had concrete parts. He also sees a sketch that was sent from the engineer to the architect showing an early conceptual diagram of the design of the frame. He saves this sketch to his local hard drive.
Matthew recalls that the specifications of the cooling tower unit itself, which were provided by the HVAC (heating, ventilation and air conditioning) subcontractor, had a large impact on the design. Nick now feels confident enough to design the new cooling tower frame by reusing the same concepts as the Bay Saint Louis cooling tower frame, as well as some of the standard details.	Nick has indicated to CoMem that he is interested in reusing this Bay Saint Louis cooling tower frame. The system highlights some related items. One of them is the cooling tower itself that is supported by the frame. Nick follows the link to this information and sees some important information about the size and weight of this cooling tower supplied by the vendor.

Table 17: Claims analysis of the activity scenario.

Situation posture	Possible pros or cons of the feature

Situation posture	Possible pros or cons of the feature
System automatically identifies relevant projects and items	+ Useful in the absence of the mentor. − Automatic relevance measure unlikely to be as effective as that provided by the mentor. − Might provide too many or too few "hits".
Zooming in and out of smaller or larger grains of the design	+ Arguably analogous to current practice: start with general notes page (whole project), select plan or section (e.g. structure), select detail (e.g. cooling tower frame). − Single 3D drawing different from idea in current practice of drawings sheets with plans, sections, and details.
Exploring the evolution of an item	+ Useful for understanding how this item was designed. − Versions captured in system might not reflect true evolution of this component. − Displaying design versions does not guarantee that the reuser will be able to observe or understand the design expertise that went into making these design decisions. − There might be too many versions, information overload.
Providing notes and messages	+ Might help reuser to understand the item and its evolution. + Will contain embedded knowledge, how and why decisions were taken. − Might be too many, information overload. − Might compromise privacy of original designers. − Liability and ownership concerns.
Highlighting related items	+ Useful in the absence of the mentor. − Automatic identification of related items unlikely to be as effective as those identified by mentor.

Table 18: Information design metaphors for CoMem.

Information	Real world metaphor	Information design ideas
An overview of the corporate memory looks like…	A map of the corporate memory	Display the entire corporate memory and allow the user to zoom in and out, or filter items
	A Windows Explorer view of the corporate memory	Use a collapsible/expandable tree view of the corporate memory
The project context (related items) of any given item looks like…	Looking at the corporate memory through a fisheye lens focused on that item	A fisheye or focus+context view: hyperbolic trees, cone trees, perspective wall, zooming and distortion
The evolution history of an item looks like…	A visual story, a comic book	Visually display the versions along with related content on a comic book-like canvas
	A flipbook	A simple listing of versions along with a display area for displaying the selected versions

Table 19: The activity scenario transformed into an information scenario.

Activity Design Scenario	Information Design Scenario
<Background on Nick, his motivations, …>	<Background on Nick, his motivations, …>
Nick gets stuck, but Matthew is not around to help. Nick decides to use the company's corporate memory system, CoMem. He identifies his current project to the system and CoMem lists some similar projects for him to look at. Nick chooses to explore the Bay Saint Louis project. He "zooms in" on the structural system for that project. With the structure in front of him, he zooms in further on the cooling tower frame.	CoMem displays an overview of the entire corporate memory in the form of a treemap. The corporate memory is hierarchical. It consists of multiple projects, each project contains multiple disciplines, and each discipline contains multiple components. Each rectangle on the treemap is color-coded according to its relevance. Nick applies some filters to reduce the number of rectangle on the treemap. Nick notices that the Bay Saint Louis cooling tower is highlighted and selects it.
Nick sees the evolution of this cooling tower frame as a series of versions. He can also see the notes and messages that were exchanged between the team members in generating these versions. He sees that the final version of the frame was steel. His current project has a concrete structure. Nick flips through the earlier versions and notices that some of them had concrete parts. He also sees a sketch that was sent from the engineer to the architect showing an early conceptual diagram of the design of the frame. He saves this sketch to his local hard drive.	The evolution appears in a separate display as a tree structure with links drawn between parent and child versions. Each version is represented by a circle, color-coded according to its level of importance and level of sharing. Any content (sketches, images, notes, CAD drawings) linked to that version is drawn as a separate node with a link to the circle representing the version. Overall, this display shows the evolution of the design of the cooling tower, a little bit like a comic book.

Activity Design Scenario	Information Design Scenario
Nick has indicated to CoMem that he is interested in reusing this Bay Saint Louis cooling tower frame. The system highlights some related items. One of them is the cooling tower itself that is supported by the frame. Nick follows the link to this information and sees some important information about the size and weight of this cooling tower supplied by the vendor.	The project context of the cooling tower appears in a separate display, again as a tree structure represented by nodes and links. Nick sees related components within the same discipline, related disciplines within the same project, and related projects in the corporate memory. Nick clicks on the HVAC discipline from the same project. Its contents appear. One of them is the cooling tower which is highlighted as a relevant item.

Table 20: Claims analysis used to guide and document information design reasoning.

Design feature	Possible pros or cons of the feature
Hierarchical information structures	+ This is how the information is captured in the ProMem system (makes implementation of CoMem easier).
	+ Hierarchies are simple and easy for users to understand.
	− Can be overly simplistic.
	− Hierarchy might not be best option for viewing related items (versus a simple list of related items).
Corporate memory is organized in a project hierarchy rather than topic hierarchy	+ This is how ProMem structures the information.
	+ Designers think back to specific projects rather than specific topics.
	+ Does not require human editor to organize information into a topic hierarchy or classification system.
	− Organizing by topic would have been useful in grouping related and similar items together.
Treemap	+ Makes full use of display space.
	+ More effective for large hierarchies than other visualizations.
	− Will be unfamiliar to most new users (+ but arguably is learnable).
Overview shows everything	+ User always starts with same initial view – will develop familiarity with the corporate memory.
	+ "False negatives" less likely.
	+ More control to the user to filter as desired.
	− Overview can be (initially) overwhelming.
	− In some cases there may be only a few items (out of tens of thousands) that are relevant. Showing everything in such cases does not make sense.
Fisheye lens project context	+ Conveys the idea of focus and context.
	− Not applied literally (i.e. the display does not visually look like a fisheye lens) and so can be confusing.

165

Design feature	Possible pros or cons of the feature
Storytelling evolution history	+ Simple, informal.
	− Storytelling metaphor can prevent it from being taken seriously.
	− There might be too many versions, or too many items attached to each version (+ can filter).

Table 21: CoMem metaphors, with emphasis on interaction design.

Interaction	Real world metaphor	Information design ideas
Looking at an overview of the corporate memory is like…	Looking at a map	Give overview appearance of a map
Finding a specific item on the overview is like…	Visually scanning the map	Label the map, allow different levels of detail and labeling to help visual search
Looking more closely at a specific item or project on the map is like…	Zooming in on a region of the map	Make the map zoomable
	(Filtering out unrelated items – not consistent with real world map metaphor, but useful)	Allow dynamic queries and filtering

Table 22: Fully detailed interaction scenario.

Interaction Scenario

As before, Matthew and Nick are working on a ten-storey hotel that has a large cooling tower unit and Nick is assigned the task of designing the frame that will support this cooling tower. They are using the ProMem system. Nick gets stuck, but Matthew is not around to help. Nick clicks on the Reuse button in ProMem, which brings up CoMem. CoMem displays a map of the entire "X Inc" corporate memory. Items on the map are color-coded according to how relevant they are to his current project. Nick uses sliders to filter out irrelevant projects, disciplines, and components from the map. Most of the rectangles in the map are now grayed out. Of the few items that remain highlighted, Nick notices the Bay Saint Louis project. However, because the corporate memory is so large, Nick cannot make out the contents of the Bay Saint Louis project. He clicks on a checkbox which causes grayed out items to disappear. The unfiltered items now take up the entire display area. Nick can see the contents of the Bay Saint Louis project much more clearly. It has a relevant Engineering discipline, and several relevant components within that discipline. He clicks on the component labeled Cooling Tower Frame.

The project context and evolution history of the Bay Saint Louis cooling tower frame appear in two separate displays. Nick examines the evolution of the frame. There are dozens of versions and Nick cannot make out the evolution clearly. He chooses to see only milestone versions of the evolution. He sees that the cooling tower frame started as a composite steel-concrete frame but was later changed into a steel frame. He sees several icons representing notes attached to the various versions. He clicks on these icons to expand them, and the fully expanded notes appear. He sees that these notes were exchanged between the architect and engineer that help to explain the change from a composite frame to a steel frame. Nick clicks on one of the versions, and a detailed view of this version appears. He finds a useful early sketch of the composite frame, which he saves to his local hard drive.

Next, Nick begins to explore the project context of the Bay Saint Louis frame. He clicks on the Engineering discipline object in the Project Context Explorer and sees that the Bay Saint Louis structural design criteria are similar to those in his current project. He notices a related component under the HVAC discipline: it is labeled Cooling Tower. This is the air conditioning unit that is supported by the frame. Nick finds a specifications sheet attached to this component. It gives him an idea of the loads for which he must now design his cooling tower.

Table 23: Analysis of the scenario.

Scenario feature	Possible pros or cons of the feature
Dynamic queries in the Overview	+ Very useful for locating reusable items. − Inconsistent with map metaphor, no real world equivalent.
Filtered items are grayed out	+ In keeping with map metaphor, "geography" of corporate memory does not change. − For very large corporate memories, it can still be difficult to see unfiltered items after filtered items are grayed out.
Filtered items disappear	+ Leaves more space for unfiltered items. − Sudden layout changes can be disorientating. − No map equivalent.
Content on Evolution History Explorer represented as clickable icons	+ Saves space. + User has control of which items to expand or minimize. − Adds complexity, can be unnecessary and irritating if user's preference is to have all items expanded.
Evolution history canvas is pan-able and zoomable	+ Removes (almost) all space constraints. + Maybe subjectively pleasing interaction (to zoom and pan). − Adds complexity, can be unnecessary for small numbers of versions.

EXPLORATION TASKS USED DURING EVALUATION
OF COMEM

Figure 73 in Chapter 11 shows Exploration Task I used in the evaluation of CoMem. The figures below show the remaining five exploration tasks used.

TASK II = PT SLAB		
ProblemIndex=130		
Reusable	**Reusable Items**	**Context Questions**
Pacific 1999		
	Pacific 1999>Structure>290-1st Floor (slabs)	What was the thickness of the slabs? (12in)
	Pacific 1999>Structure>287-2nd Floor (slabs)	Did you see the slab calculations? (yes)
	Pacific 1999>Structure>288-3rd Floor (slabs)	
Pacific 2001		
	Pacific 2001>Arch>411-All Level Plan	What depth will the MEP take? (90cm)
	Pacific 2001>Arch>410-Slab Penetrations	Were the slabs penetrated? (No)
	Pacific 2001>Const>482-PT Slabs	Which program was used to design the slabs? (Floor)
	Pacific 2001>Struct>444-PT Slabs (PT slabs)	What was main reason for choice of PT slabs? (deflection, advice of mentors)
	Pacific 2001>Struct>465-1st Floor Slabs	What worked together with the slab to reduce deflection? (the columns)
	Pacific 2001>Struct>455-Overall Structural Performance	
Wave 2001		
	Wave 2001>ConstSeq>401-Slabs F1	What should the engineer check? (Punching shear)
	Wave 2001>ConstSeq>402-Slabs F2	
	Wave 2001>ConstSeq>403-Slabs F3	
	Wave 2001>Struct>360-Slabs (slabs)	
	Wave 2001>Struct>359-PT slabs (slabs)	
Coral 2002		
	Coral 2002>Structure>919-Typical Slab	What change did the egineer make to make the slab work? (Move some columns)
	Coral 2002>Structure>937-Slabs1 (slabs)	What are typical min/max moments and deflections in the slab? (check that they can point to this information)
	Coral 2002>Structure>841-Slabs1 (slabs)	What is the typical depth of the slab? (12in)
	Coral 2002>Structure>1014-Slabs2 (slabs)	
	Coral 2002>Structure>1017-Slabs1 (slabs)	
	Coral 2002>Typical Details>1031-Column Beam Section	
Ridge 2002		
	X	
Island 2002		
	X	
Bay Saint Louis		
	BSL>Structure>31-Ballroom (slab)	
	BSL>Structure>33-Central plant (slab)	
Grand Californian Hotel		
	GCH>Structure>45	
	GCH>Structure>48	
	GCH>Structure>50	
	GCH>Structure>51	
	GCH>Structure>54	
	GCH>Structure>60	
San Rafael Bridge Retrofit		
	X	
Imaginary Hotel		
	X	

Figure 85: A sample exploration task, where the user is searching for reusable items related to post tensioned slabs.

171

TASK III = SHEAR WALLS ProblemIndex=129		
Reusable	**Reusable Items**	**Context Questions**
Pacific 1999		
	Pacific 1999>Struct>305-1st floor exterior (shear wall)	Did you see shear wall calcs? (yes)
	Pacific 1999>Struct>302-2nd floor exterior (shear wall)	Why were the shear walls so thick? (architectural concept)
	Pacific 1999>Struct>303-3rd floor exterior (shear wall)	What other building component will be made to match the material of the shear walls? (Auditorium walls)
	Pacific 1999>Struct>296-1st floor interior (shear wall)	
	Pacific 1999>Struct>297-1st floor interior (shear wall)	
	Pacific 1999>Struct>294-3rd floor interior (shear wall)	
	Pacific 1999>Arch>279-Auditorium (ExteriorWalls)	
Pacific 2001		
	Pacific 2001>Arch>408-Sloping walls (wall)	How thick are the shear walls? (6 in)
	Pacific 2001>Const>480-1st floor shear walls	Why only 6in? (Because more shear walls than necessary are used)
	Pacific 2001>Const>478-2nd floor shear walls	Which connection was the team worried about? (Collector beam to shear wall)
	Pacific 2001>Const>484-3rd floor shear walls	What helps the shear walls resist lateral loads on the third floor? (Lateral columns)
	Pacific 2001>Struct>452-1st floor shear walls	Where the two halves of the building connected structurally? (No, at least originally)
	Pacific 2001>Struct>463-2nd floor lateral columns	Why are the thrid floor lateral columns important? (Earthquake forces)
	Pacific 2001>Struct>461-1st floor core	
	Pacific 2001>Struct>468-3rd floor gravity (columns)	
	Pacific 2001>Struct>466-1st floor collector (beams)	
	Pacific 2001>Struct>470-3rd floor core	
	Pacific 2001>Struct>471-3rd floor lateral columns	
	Pacific 2001>Struct>448-Typical Details (Shear walls)	
	Pacific 2001>Struct>457 Lateral cliff	
	Pacific 2001>Struct>458 Lateral cliff	
Wave 2001		
	Wave 2001>Struct>358-Shear walls (shear walls)	
Coral 2002		
	X	
Ridge 2002		
	Ridge 2002>Arch>836-Auditorium Shear Walls	What finish was the team considering for the auditorium shear walls? (stone finish)
	Ridge 2002>Struct>496-Shear core (shearwalls)	Why did the mentor advice them to reconsider this? (extra load) What was the mentor's feedback on the shear walls? (might not be sufficient)
Island 2002		
	Island 2002>Walls>1103-Shear walls (shear walls)	Why was the final version of the shear walls chosen? (least displacement)
	Island 2002>Struct>1078-Shear walls (shear)	
	Island 2002>Struct>1075-Shear walls (footings)	
Bay Saint Louis		
	X	
Grand Californian Hotel		
	X	
San Rafael Bridge Retrofit		
	X	
Imaginary Hotel		
	X	

Figure 86: A sample exploration task, where the user is searching for reusable items related to shear walls.

TASK IV = ATRIUM ProblemIndex=135		
Reusable	**Reusable Items**	**Context Questions**
Pacific 1999	Pacific 1999>Spaces>241-atrium1 (atrium)	What two suppliers were considered for this atrium? (InKan and EFCO) What did the architect consider when designing this atrium? (lighting)
Pacific 2001	Pacific 2001>Arch>412-Atrium	What was the architectural rationale is having an atrium? (to inspire curiosity, engeering building) What is the live load in the atrium? (100 psf) What materials were used in the building? (concrete, wood, sheet metal)
Wave 2001	X	
Coral 2002	Coral 2002>Arch>1035-Auditorium (spaces)	What two architectural concepts were considered for the atrium? (triangular and central core) Can you describe the structural system proposed by the engineer? (square bays, braced frames around perimeter, shear walls around core) Why did the engineer oppose the atrium? (wastes space, need to excavate, which is expensive)
Ridge 2002	X	
Island 2002	X	
Bay Saint Louis	X	
Grand Californian Hotel	GCH>Arch>88-Atrium Lobby Gridline1 (Faux Timber Truss) GCH>Arch>85-Atrium Lobby Gridline2 (Faux Timber Truss) GCH>Arch>86-Atrium Lobby Gridline3 (Faux Timber Truss) GCH>Arch>91-Atrium Lobby Gridline4 (Faux Timber Truss) GCH>Struct>53-Atrium Lobby North (King post truss) GCH>Struct>58-Atrium Lobby Center (King post truss) GCH>Struct>52-Atrium Lobby South (King post truss) GCH>Site>63-Atrium Lobby (hotel)	
San Rafael Bridge Retrofit	X	
Imaginary Hotel	X	

Figure 87: A sample exploration task, where the user is searching for reusable items related to atriums.

173

TASK V = ELEVATOR		
ProblemIndex=141		
Reusable	**Reusable Items**	**Context Questions**
Pacific 1999		
	Pacific 1999>spaces>237-elevator	Will the elevator shaft be continuous for all three floors? (yes)
	Pacific 1999>Struct>296-1st floor interior shear wall	Will there be shear walls, how many? (two according to note, three in CAD model)
	Pacific 1999>Struct>297-2nd floor interior shear wall	
Pacific 2001		
	Pacific 2001>Arch>417-Elevator	Why did the architect find the elevator difficult to locate? (because the structure had to be considered)
	Pacific 2001>Arch>431-Structural grid	Which rooms are located near the elevator on the 3rd floor? (seminar room, some faculty offices)
	Pacific 2001>Arch>436-3rd floor seminar room	In what sequence will the elevator shaft be installed? (all at once, at the same time as the first floor)
	Pacific 2001>Arch>441-3rd floor faculty offices	What transfers load to the shear walls? (collector beams)
	Pacific 2001>Const>477-Elevator shaft	What problem did the asymmetry of the building cause? (large torsional loads)
	Pacific 2001>Const>486-1st floor sequence	How did the lateral system differ from the 1st to the 3rd floor? (less shear walls, more lateral columns)
	Pacific 2001>Struct>461-1st floor core	
	Pacific 2001>Struct>466-2nd floor core	
	Pacific 2001>Struct>470-3rd floor core	
Wave 2001		
	X	
Coral 2002		
	Coral 2002>Struct>928-Elevator wall	What building element will this wall support? (stairs)
	Coral 2002>Arch>1035-Auditorium (spaces)	Can you describe the architectural concept of the elevator? (atrium with central core of elevators and stairs provides the means of vertical movement. From this central circulation core, a series of catwalks would lead you to the rooms on the two wings of the building)
Ridge 2002		
	Ridge 2002>Costs>558-Ground MEP	Why did the CM erect the elevator shaft early? (to allow time to place elevator...)
	Ridge 2002>Costs>536-First MEP	Why did the architect not want elevator near the auditorium? (some circulation reason, see DF hyperlink in last version of Design Issues)
	Ridge 2002>Costs>538-Second MEP	
	Ridge 2002>Arch>834-Design Issues	
Island 2002		
	Island 2002>Arch>1114-Floorplans (floorplans)	What else is grouped with the elevator in the building core? (stairs and bathrooms)
	Island 2002>Arch>1115-Vertical circulation (design issues)	What additional vertical circulation did they have (besides stairs and elevator)? (ramp)
	Island 2002>Finish work>1110-Elevator (doors)	Why was the ramp eliminated? (not needed, taking up space, expensive)
		What kind of elevator did the team consider? (hydraulic)
		What are the pros and cons of this type of elevator? (see note attached to 1110)
Bay Saint Louis		
	X	
Grand Californian Hotel		
	X	
San Rafael Bridge Retrofit		
	X	
Imaginary Hotel		
	X	

Figure 88: A sample exploration task, where the user is searching for reusable items related to elevators.

174

TASK VI = HVAC ProblemIndex=142		
Reusable	**Reusable Items**	**Context Questions**
Pacific 1999	X	
Pacific 2001	Pacific 2001>Const>492-HVAC considerations	What kind of HVAC system were they considering? (fuel cells)
	Pacific 2001>Const>487-Fuel cells	Where will they put the HVAC system? (below the auditorium)
	Pacific 2001>Arch>430-Auditorium	Can you explain how a fuel cell works? (like a battery, but with fuel constantly fed to cell. Fuel is converted directly to electricity)
	Pacific 2001>Arch>433-Storage	Did you locate the storage room? (yes)
Wave 2001	Wave 2001>MEP>404-Air handling unit	Where will the air-handling unit go? (on the roof, inside triangular truss)
	Wave 2001>MEP>405-Layout (duct)	What impact will this have on the structure? (columns below will be larger)
	Wave 2001>Struct>363-Roof structure	
	Wave 2001>Arch>366-Roof (roof)	
Coral 2002	Coral 2002>Const>1070-HVAC (hvac)	
Ridge 2002	Ridge2002>costs>748-HVAC (mep)	What kind of system did they choose? (air-cooled system)
	Ridge2002>arch>835-HVAC design issues	Why? (lighter, cheaper, less maintenance)
Island 2002	Island 2002>Finish work>1111-Under-floor (HVAC)	What kind of system did they consider? (under-floor)
	Island 2002>Slab>any slab	What are the pros and cons of this system? (any answer from the DF discussion or notes) What size ducts? (16" by 22")
Bay Saint Louis	BSL>HVAC>2-Primary (cooling unit)	What supplier did they use for the cooling unit? (Marley)
	BSL>Structure>any cooling tower frame	
Grand Californian Hotel	X	
San Rafael Bridge Retrofit	X	
Imaginary Hotel	X	

Figure 89: A sample exploration task, where the user is searching for reusable items related to HVAC.

INSTRUCTIONS TO PARTICIPANTS

To reduce variability in understanding the interfaces and tasks being tested between different participants, a set of instructions was prepared and read almost verbatim to each participant at the beginning of the test. These instructions are shown below.

Objective of Experiment
I am trying to evaluate and compare three prototypes for exploring a corporate memory of content from previous projects. I am not trying to assess your abilities, but to assess these prototypes.

Procedure
I will ask you to complete various tasks using three various prototypes. These tasks have to do with finding and understanding content from previous projects, so you will not have to do any actual design work, all you have to do is use the prototype systems to explore and retrieve content from the database. I tried to make the tasks and the data as realistic as possible, but I realize that they are not entirely convincing. However I ask you to suspend your disbelief and play along!

The Data
Imagine that you work for a company that keeps a large archive of all the projects it has worked on in the past. The archive is organized hierarchically: multiple projects, each project containing multiple building subsystems, each subsystem containing multiple components. The system is able to detect what project you (the user) are currently working on, and highlight relevant items from the archive for you to reuse. Note that the database can be quite sparse in some places. Projects and disciplines do not contain any data.

Interface 1: CoMem
This interface uses a treemap, where the hierarchy is represented as series of nested rectangles, like this… [Here the test conductor makes a sketch of a treemap being constructed.] The system can work in two modes: either using white rectangles or colored rectangles that indicate the relevance of each item to your current project.

Interface 2: Outline Tree
This is an interface you might be familiar with. It is similar to Windows Explorer, which is used to browse folders and files on a computer hard drive. The system can

work in two modes: either using generic icons for folders and files, or using colored icons which indicate the relevance of each item to your current project. When you click on an item, you will get a display of the item, plus a table of versions...

Interface 3: Hit List
Like Google... Works in two modes, either ranked by relevance to your task, or sorted alphabetically. You can type in keywords to filter. If you type in multiple keywords, they will be OR.

Tasks
Retrieval: I will ask you to find a specific item. When you think you are done let me know. I will time how long it takes you.

Exploration: I will give you a scenario and give you as much time as you want to find all the reusable information you can from database. When you are done, I will measure how much time you took, and then ask you some questions about those items that you retrieved. Please make a list of all the items, grouped by project. For some projects, there will be lots of content in the database for you to explore. In others, there will be little content. In both cases, include the items in your list, and if there is content, explore it, i.e. list even if no useful content. Be as inclusive as possible. Not only directly but also indirectly related. Give example.

Before we start, you have some time to play with the three prototypes.

REFERENCES

Ackerman M. S., 1994. "Augmenting the organizational memory: A field study of Answer Garden", Proceedings of the ACM Conference on Computer-Supported Cooperative Work Conference (CSCW), pages 243-252.

Ahmed S., Blessing L., and Wallace K., 1999. "The relationships between data, information and knowledge based on a preliminary study of engineering designers", Proceeding of the Eleventh International Conference on Design Theory and Methodology (DTM), ASME Design Engineering Technical Conferences (DETC).

Allen T., 1977. *Managing the Flow of Technology*, MIT Press, Cambridge, MA.

Alonso O. and Frakes W. B., 2000. "Visualization of reusable software assets", Lecture Notes in Computer Science, Proceedings of the Sixth International Conference on Software Reuse (ICSR), Advances in Software Reusability, (Frakes W. B., Ed.), Springer-Verlag, Vienna, Austria, pages 251-265.

Altmeyer J. and Shürmann B., 1996. "On design formalization and retrieval of reuse candidates", Proceedings of the Fourth International Conference on Artificial Intelligence in Design (AID), pages 231-250.

Arias E., Eden H., and Fischer G., 1997. "Enhancing communication, facilitating shared understanding, and creating better artifacts by integrating physical and computational media for design", Symposium on Designing Interactive Systems (DIS), Proceedings of the Conference on Designing Interactive Systems: Processes, Practices, Methods, and Techniques.

Baeza-Yates R. and Ribeiro-Neto B., 1999. *Modern Information Retrieval*, Addison-Wesley, Harlow, UK.

Baudin C., Underwood J., and Baya V., 1993. "Using device models to facilitate the retrieval of multimedia design information", Proceedings of Thirteenth International Joint Conference on Artificial Intelligence (IJCAI), pages 1237-1243.

Bederson B. B., Hollan J. D., Stewart J., Rogers D., Druin A., and Vick D., 1996. "A zooming web browser", SPIE Multimedia Computing and Networking, Volume 2667, pages 260-271.

Bederson B., Meyer J., and Good L., 2000. "Jazz: An extensible zoomable user interface graphics toolkit in Java", Proceedings of the Thirteenth Annual ACM Symposium on User Interface and Software Technology (UIST), pages 171-180.

Bhatta S., Goel A., and Prabhakar S., 1994. "Innovation in analogical design: A model-based approach", Proceedings of the Third International Conference on Artificial Intelligence in Design (AID), pages 57-74.

Bilgic T. and Fox M. S., 1996. "Case-based retrieval of engineering design cases: Context as constraints", Proceedings of the Fourth International Conference on Artificial Intelligence in Design (AID), pages 269-288.

Blomberg J., Giacomi J., Mosher A., and Swenton-Wall P., 1993. "Ethnographic field methods and their relation to design", *Participatory Design: Principles and Practices*, (Schuler D., Ed.), Lawrence Erlbaum Associates, New Jersey, pages 123-155.

Brin S. and Page L., 1998. "The anatomy of a large-scale hypertextual web search engine", Computer Networks and ISDN Systems, Volume 30, Issue 1-7, pages 107-117.

Brooke J., 1996. "SUS: A quick and dirty usability scale", *Usability Evaluation in Industry*, (Jordan P. W., Thomas B., Weerdmeester B. A., and McClelland I. L., Eds.), Taylor and Francis, London, UK, pages 189-194.

Brown J. S. and Duguid P., 2000. *The Social Life of Information*, Harvard Business School Publishing, Boston, MA.

Bruls D. M., Huizing K., and van Wijk J. J., 1999. "Squarified treemaps", Data Visualization 2000, Proceedings of the Second Joint Visualization Symposium organized by the Eurographics and the IEEE Computer Society Technical Committee on Visualization and Graphics (TCVG), (de Leeuw W. and van Liere R., Eds.), Springer-Verlag, Vienna, Austria, pages 33-42.

Bucciarelli L. L., 1994. *Designing Engineers*, MIT Press, Cambridge, MA.

Burleson W. and Selker T., 2002. "Creativity and interface", Communications of the ACM, Special Issue on Creativity and Interface, Volume 45, Issue 10, pages 88-90.

Caldas C. H., Soibelman L., and Han J., 2002. "Automated classification of construction project documents", ASCE Journal of Computing in Civil Engineering, Volume 16, Number 4, pages 234-243.

Card S. K., Mackinlay J. D., and Shneiderman B., 1999. *Readings in Information Visualization: Using Vision to Think*, Morgan Kaufmann Publishers, San Francisco, CA.

Carroll J. M., 2000. *Making Use: Scenario-Based Design of Human-Computer Interactions*, MIT Press, Cambridge, MA.

Catledge L. D. and Pitkow J. E., 1995. "Characterizing browsing strategies in the World-Wide Web", Computer Networks and ISDN Systems, Volume 27, Issue 6, pages 1065-1073.

Chung P. and Goodwin R., 1994. "Representing design history", Proceedings of the Second International Conference on Artificial Intelligence in Design (AID), pages 735-751.

179

Combs T. and Bederson B., 1999. "Does zooming improve image browsing?", Proceedings of the Fourth ACM International Conference on Digital Libraries, pages 130-137.

Cross N., 1989. *Engineering Design Methods*, John Wiley and Sons, Chichester, NY.

Culley S. J. and Theobald G., 1997. "Dealing with standard components for knowledge intensive CAD", *Knowledge Intensive CAD*, (Mäntylä M., Finger S., and Tomiyama T., Eds.), Volume II, Chapman and Hall, London, UK, pages 235-255.

Culley S. J., 1998. "Design reuse of standard parts", Proceedings of the Engineering Design Conference on Design Reuse, pages 77-88.

Culley S. J., 1999. "Classification approaches for standard parts to aid design reuse", Proceedings of the Institution of Mechanical Engineers, Part B, Journal of Engineering Manufacture, Volume 213, Issue 2, pages 203-207.

Danielson D. R., 2002. "Transitional volatility in Web navigation: Usability metrics and user behavior", Master of Science Thesis, Stanford University.

Darken R. P. and Sibert J. L., 1993. "A toolkit for navigation in virtual environment", Proceedings of the Sixth Annual ACM Symposium on User Interface Software and Technology (UIST), pages 157-165.

Davis J. E., (Ed.), 2002. *Stories of Change: Narrative and Social Movements*, State University of New York Press, Albany, NY.

Deerwester S., Dumais S. T., Furnas G. W., Landauer T. K, and Harshman R., 1990. "Indexing by Latent Semantic Analysis", Journal of the American Society for Information Science, Volume 41, Issue 6, pages 391-407.

Demian P. and Fruchter R., 2005. "Measuring relevance in support of design reuse from archives of building product models", ASCE Journal of Computing in Civil Engineering, volume 29, issue 2, pp. 119-136.

Demian, P. and Fruchter, R., 2006a. "A Methodology for Usability Evaluation of Corporate Memory Design Reuse Systems." ASCE Journal of Computing in Civil Engineering, Volume 20, Issue 6, pp. 377-389.

Demian, P. and Fruchter, R., 2006b. "Finding and Understanding Reusable Designs from Large Hierarchical Repositories." Information Visualization Journal, Volume 5, Number 1, pp. 28-46.

Demian, P. and Fruchter, R., 2006c. "An Ethnographic Study of Design Knowledge Reuse in the Architecture, Engineering and Construction Industry" Journal of Research in Engineering Design, volume 16, number 4, pp. 184-195.

Dingsøyr T., 1998. "Retrieval of cases by using a Bayesian network", Papers from the AAAI Workshop on Case-Based Reasoning Integrations, pages 50-54.

Domeshek E. and Kolodner J., 1993. "Finding the points of large cases", Artificial Intelligence for Engineering Design, Analysis and Manufacturing (AI EDAM), Volume 7, Issue 2, pages 87-96.

Dorst K., 1997. "Describing design: A comparison of paradigms", Doctoral Thesis, Delft University of Technology, The Netherlands.

Duffy S. M., Duffy A. H. B., and MacCallum K. J., 1995. "A design reuse process model", Proceedings of the Tenth International Conference on Engineering Design (ICED), pages 490-495.

Dumais S. T., 1991. "Improving the retrieval of information from external sources", Behavior Research Methods, Instruments, and Computers, Volume 23, Number 2, pages 229-236.

Duncan R., 1999. "Toward a theory of comic book communication", Presented at the 85th Annual Convention of the National Communication Association (NCA). Also available from the Academic Forum Online, (Fudge K., Ed.), 1999-00, Number 17, Henderson State University, Arkadelphia, AR.

Eastman C. M., 1999. *Building Product Models: Computer Environments Supporting Design and Construction*, CRC Press, Boca Raton, FL.

Edson E., 2001. "Bibliographic essay: History of cartography", CHOICE: Current Reviews for Academic Libraries, July/August 2001, Volume 38, Number 11/12, pages 1899-1909.

Erickson T., 1995. "Notes on design practice: stories and prototypes as catalysts for communication", *Scenario-based design: Envisioning work and technology in system development*, (Carroll J. M., Ed.), John Wiley and Sons, New York, NY, pages 37-59.

Ferguson E. S., 1992. *Engineering and the Mind's Eye*. MIT Press, Cambridge, MA.

Finger S., 1998. "Design reuse and design research – Keynote paper", Proceedings of the Engineering Design Conference 1998: Design Reuse, (Sivaloganathan S. and Shahin T. M. M., Eds.), pages 3-10.

Fiore A. and Smith M. A., 2001. "Treemap visualizations of newsgroups", Technical Report, Microsoft Research, Microsoft Corporation, Redmond, WA.

Fruchter R., Saxena K., Breidenthal M. and Demian P., 2007. "Collaborative design exploration in an interactive workspace." AI EDAM Special Issue: Support for Design Teams, Volume 21, Issue 03, June 2007, pp 279-293

Fruchter R. and Demian P., 2002(a). "CoMem: Design knowledge reuse from a corporate memory", ASCE Proceedings of the Ninth International Conference on Computing in Civil and Building Engineering (ICCCBE-IX), Volume 2, pages 1145-1150.

181

Fruchter R. and Demian P., 2002(b). "Corporate memory in action", Proceedings of the ASCE International Workshop on Information Technology in Civil Engineering, Computing in Civil Engineering, pages 90-102.

Fruchter R. and Demian P., 2002(c). "Knowledge management for reuse", Proceedings of the CIB W78 Conference, Distributing Knowledge in Building, Volume 1, pages 93-100.

Fruchter R. and Demian P., 2002. "CoMem: Designing an interaction experience for reuse of rich contextual knowledge from a corporate memory", Artificial Intelligence for Engineering Design, Analysis and Manufacturing (AI EDAM), Volume 16, Issue 3, pages 127–147.

Fruchter R. and Demian P., 2005. "Corporate memory", *Knowledge Management in Construction*, (Anumba C. J., Ed.), Thomas Telford, London, UK.

Fruchter R., 1996. "Conceptual, collaborative building design through shared graphics", IEEE Expert: Intelligent Systems, AI in Civil and Structural Engineering, Volume 11, Number 3, pages 33-41.

Fruchter R., Clayton M. J., Krawinkler H., Kunz J., and Teicholz P., 1996. "Interdisciplinary communication medium for collaborative conceptual building design", Journal of Advances in Engineering Software, Computing in Civil and Structural Engineering, Volume 25, Issues 2-3, pages 89-101.

Fruchter R., Reiner K., Leifer L., and Toye G., 1998. "VisionManager: A computer environment for design evolution capture", Journal of Concurrent Engineering: Research and Applications (CERA), Volume 6, Number 1, pages 71-84.

Fulford R., 1999. *The Triumph of Narrative: Storytelling in the Age of Mass Culture*, House of Anansi Press, Toronto, Canada.

Furnas G. W. and Zacks J., 1994. "Multitrees: Enriching and reusing hierarchical structure", Proceedings of the ACM Computer Human Interaction (CHI) Conference, Human Factors in Computing Systems, pages 330-336.

Furnas G. W., 1981. "The FISHEYE view: A new look at structured files", *Readings in Information Visualization: Using Vision to Think*, 1999, (Card S. K., Mackinlay J. D., and Shneiderman B., Eds.), Morgan Kaufmann Publishers, San Francisco, CA, pages 312-330.

Garcia A. C. B., Carretti C. E., Ferraz I. N., and Bentes C., 2002. "Sharing design perspectives through storytelling", Artificial Intelligence for Engineering Design, Analysis and Manufacturing (AI EDAM), Volume 16, Issue 3, pages 229-241.

Gerbé O., 1997. "Conceptual graphs for corporate knowledge repositories", Lecture Notes in Artificial Intelligence, Proceedings of the Fifth International Conference on Conceptual Structures (ICCS), Conceptual Structures: Fulfilling Peirce's Dream, (Lukose D., Delugach H., Keeler M., Searle L., and Sowa J., Eds.), Springer-Verlag, Vienna, Austria, pages 474-488.

Gero J. S., 1990. "Design prototypes: A knowledge representation schema for design", AI Magazine, Volume 11, Number 4, pages 26-36.

Gershon N. and Page W., 2001. "What storytelling can do for information visualization", Communications of the ACM, Volume 44, Number 8, pages 31-37.

Gerstberger P. G. and Allen T. J., 1968. "Criteria used by research and development engineers in the selection of an information source", Journal of Applied Psychology, Volume 52, Number 4, pages 272-279.

Golub G. and Reinsch C., 1971. *Handbook for Automatic Computation II, Linear Algebra*, Springer-Verlag, New York.

Grudin J., 2001. "Desituating action: Digital representation of context", Human-Computer Interaction, Special Issue on Context-Aware Computing, Volume 16, pages 269-296.

Harley J. B., 2001. *The New Nature of Maps: Essays in the History of Cartography*, (Laxton P., Ed.), The John Hopkins University Press, Baltimore, MD.

Hill A., Song S., Dong A., and Agogino A., 2001. "Identifying shared understanding in design using document analysis", Proceeding of the Thirteenth International Conference on Design Theory and Methodology (DTM), ASME Design Engineering Technical Conferences (DETC).

Hollingshead A. B., 1998. "Retrieval processes in transactive memory systems", Journal of Personality and Social Psychology, Volume 74, Number 3, pages 659-671.

Jerding D. F. and Stasko J. T., 1994. "Using visualization to foster object-oriented program understanding", Georgia Institute of Technology, Atlanta, GA, Graphics, Visualization, and Usability Center (GVU), Technical Report GIT-GVU-94-33.

Jerding D. F., Stasko J. T., and Ball T., 1997. "Visualizing interactions in program executions", Proceedings of the Nineteenth International Conference on Software Engineering (ICSE), pages 360-370.

Johanson B., Fox A., and Winograd T., 2002. "The interactive workspaces project: Experiences with ubiquitous computing rooms", IEEE Pervasive Computing, Volume 1, Number 2, pages 67-74.

Johnson B. and Shneiderman B., 1991. "Treemaps: A space-filling approach to the visualization of hierarchical information structures", *Readings in Information Visualization: Using Vision to Think*, 1999, (Card S. K., Mackinlay J. D., and Shneiderman B., Eds.), Morgan Kaufmann Publishers, San Francisco, CA, pages 152-159.

Jung Y., Park H., and Du D., 2000. "An effective term-weighting scheme for information retrieval", University of Minnesota, Minneapolis, MN, Department of Computer Science and Engineering, Technical Report 00-008.

Kazman R. and Carrière S. J., 1998. "View extraction and view fusion in architectural understanding", Proceedings of the Fifth International Conference on Software Reuse (ICSR), pages 290-299.

Kleinberg J. M., 1999. "Authoritative sources in a hyperlinked environment", Journal of the ACM, Volume 46, Number 5, pages 604-632.

Koike H. and Yoshihara H., 1993. "Fractal approaches for visualizing huge hierarchies", Proceedings of the IEEE Symposium on Visual Languages, pages 55-60.

Kuffner T. A. and Ullman D. G., 1990. "The information requests of mechanical design engineers", Proceedings of the Second International Conference on Design Theory and Methodology (DTM), ASME Design Engineering Technical Conferences (DETC), pages 167-174.

Lakin F., Wambaug H., Leifer L., Cannon D., and Sivard C., 1989. "The electronic design notebook: Performing medium and processing medium", Visual Computer: International Journal of Computer Graphics, Volume 5, Number 4, pages 214-226.

Lamping J. and Rao R., 1995. "The hyperbolic browser: A Focus+Context technique for visualizing large hierarchies", *Readings in Information Visualization: Using Vision to Think*, 1999, (Card S. K., Mackinlay J. D., and Shneiderman B., Eds.), Morgan Kaufmann Publishers, San Francisco, CA, pages 382-408.

Landauer T. K. and Dumais S. T., 1997. "A solution to Plato's problem: The latent semantic analysis theory of acquisition, induction and representation of knowledge", Psychological Review, Volume 104, Number 2, pages 211-240.

Lee M. L., Yang L. H., Hsu W., and Yang X., 2002. "XClust: Clustering XML schemas for effective integration", Proceedings of the Eleventh International Conference on Information and Knowledge Management (CIKM), pages 292-299.

Leifer L., 1997. Design Project Lab. (ME310), Stanford University, Course Notes.

Lloyd P., Busby J., and Deasley P., 1998. "Reuse to overuse – A problem of fixation?", Proceedings of the Engineering Design Conference on Design Reuse, pages 457-466.

Luth G. P., 1991. "Representation and reasoning for integrated structural design", Doctoral Thesis, Stanford University.

Maher M. L. and Gómez de Silva Garza A., 1996. "Developing case-based reasoning for structural design", IEEE Expert: Intelligent Systems, AI in Civil and Structural Engineering,, Volume 11, Number 3, pages 42-52.

Maher M. L., 1997. "CASECAD and CADSYN: Implementing case retrieval and case adaptation", *Issues and Applications of Case-Based Reasoning in Design*, (Maher M. L. and Pu P., Eds.), Lawrence Erlbaum Associates, Mahwah, NJ, pages 161-185.

McCall R., 1987. "PHIBIS: Procedurally Hierarchical Issue-Based Information Systems", Proceedings of the ASME Conference on Architecture at the International Congress on Planning and Design Theory, pages 17-22.

McCloud S., 1993. *Understanding Comics: The Invisible Art*, Kitchen Sink Press, Northampton, MA.

Nardi B. A., Whittaker S., and Schwarz H., 2000. "It's not what you know, it's who you know: Work in the information age", First Monday: Peer-Reviewed Journal of the Internet, Volume 5, Number 5.

Nardi B. A., Whittaker S., Isaacs E., Creech M., Johnson J., and Hainsworth J., 2002. "ContactMap: Integrating communication and information through visualizing personal social networks", Communications of the ACM, Volume 45, Number 4, pages 89-95.

Nelson T. H., 1990. "The right way to think about software design", *The Art of Human-Computer Interface Design*, (Laurel B., Ed.), Addison-Wesley, Reading, MA, pages 235-244.

Nielsen J., 1993. *Usability Engineering*, Academic Press, San Diego, CA.

Nielsen J., 1999. "User interface directions for the Web", Communications of the ACM, Volume 42, Number 1, pages 65-72.

Nielsen J., 2000. *Designing Web Usability: The Practice of Simplicity*, New Riders, Indianapolis, IN.

Perlin K. and Fox D., 1993. "Pad: An alternative approach to the computer interface", Proceedings of the Twentieth ACM International Conference on Computer Graphics and Interactive Techniques (ACM SIGGRAPH), pages 57-64.

Pirolli P. and Card S., 1999. "Information foraging", Psychological Review, Volume 106, Number 4, pages 643-675.

Pirolli P., Card S. K., and Van Der Wege M. M., 2000. "The effect of information scent on searching information visualizations of large tree structures", Proceedings of the ACM Working Conference on Advanced Visual Interfaces (AVI), pages 161-172.

Pisupati C., Wolff L., Mitzner W., and Zerhouni E., 1996. "Geometric tree matching with applications to 3D lung structures", Proceedings of the Twelfth ACM Annual Symposium on Computational Geometry, pages 419-420.

Polanyi M., 1966. *The Tacit Dimension*, Doubleday, Garden City, NY.

Pollock J. L. and Cruz J., 1999. *Contemporary Theories of Knowledge*, Rowman and Littlefield Publishers, Lanham, MD.

Popova M., Johansson P., and Lindgren H., 2002. "An integrated platform for case-based design", Proceedings of the CIB W78 Conference, Distributing Knowledge in Building, Volume 2, pages 99-106.

Rasmussen J., 1990. "Mental models and the control of action in complex environments", *Mental Models and Human-Computer Interaction 1*, Selected Papers of the Sixth Interdisciplinary Workshop in Informatics and Psychology, North-Holland, New York, NY, pages 41-69.

Regli W. C., Hu X., Atwood M., and Sun W., 2000. "A survey of design rationale systems: Approaches, representation, capture and retrieval", Engineering with Computers, Volume 16, Numbers 3-4, Springer-Verlag, Vienna, Austria, pages 209-235.

Rehder B., Schreiner M. E., Wolfe M. B. W., Laham D., Landauer T. K., and Kintsch W., 1998. "Using Latent Semantic Analysis to assess knowledge: Some technical considerations", Discourse Processes, The Official Journal of the Society for Text and Discourse, Special Issue: Quantitative Approaches to Semantics Knowledge Representations, Volume 25, Numbers 2 and 3, pages 337-354.

Reiner K. and Fruchter R., 2000. "Project memory capture in globally distributed facility design", ASCE Proceedings of the Eighth International Conference on Computing in Civil and Building Engineering (ICCCBE-VIII), Volume 2, pages 820-827.

Retkowsky F., 1998. "Software reuse from an external memory: The cognitive issues of support tools", Proceedings of the Tenth Workshop on Psychology of Programming Interest Group (PPIG).

Robertson G. G., Mackinlay J. D., and Card S. K., 1991. "Cone trees: Animated 3D visualizations of hierarchical information", Proceedings of the ACM Computer Human Interaction (CHI) Conference, Human Factors in Computing Systems, pages 189-194.

Rosson M. B. and Carroll J. M., 2001. *Usability Engineering: Scenario-Based Development of Human Computer Interaction*, Morgan Kaufmann, San Francisco, CA.

Salton G. and Buckley C., 1988. "Term-weighting approaches in automatic text retrieval", Information Processing and Management, An International Journal, Volume 24, Number 5, pages 513-523.

Salton G., Allan J., Buckley C., and Singhal A., 1995. "Automatic analysis, theme generation, and summarization of machine-readable texts", *Readings in Information Visualization: Using Vision to Think*, 1999, (Card S. K., Mackinlay J. D., and Shneiderman B., Eds.), Morgan Kaufmann Publishers, San Francisco, CA, pages 413-418.

Schank R. C., 1990. *Tell Me a Story: A New Look at Real and Artificial Memory*, Scribner, New York, NY.

Schön D. A., 1983. *The Reflective Practitioner: How Professionals Think in Action*, Basic Books, New York, NY.

Shahin T. M. M., Sivaloganathan S., and Gilliver R., 1997. "Automation of feature-based modelling and finite element analysis for optimal design", Proceedings of the Eleventh International Conference on Engineering Design (ICED).

Shneiderman B. and Wattenberg M., 2001. "Ordered treemap layouts", Proceedings of the IEEE Symposium on Information Visualization, pages 73-78.

Shneiderman B., 1994. "Dynamic queries for visual information seeking", *Readings in Information Visualization: Using Vision to Think*, 1999, (Card S. K., Mackinlay J. D., and Shneiderman B., Eds.), Morgan Kaufmann Publishers, San Francisco, CA, pages 236-243.

Shneiderman B., 1999. "Supporting creativity with advanced information-abundant user interfaces", University of Maryland, College Park, MD, The Institute for Systems Research (ISR), Technical Report 1999-73. Also available from *Frontiers in Human-Centred Computing, Online Communities and Virtual Environment*, 2001, (Earnshaw R., Guedj R., Van Dam A., and Vince J., Eds.), Springer-Verlag, London, UK, pages 469-480.

Shneiderman B., 2002. *Leonardo's Laptop: Human Needs and the New Computing Technologies*, MIT Press, Cambridge, MA.

Simmons A., 2001. *The Story Factor: Secrets of Influence from the Art of Storytelling*, Perseus Publishing, Cambridge, MA.

Simoff S. J. and Maher M. L., 1998. "Data mining in hypermedia case libraries", Proceedings of the Fifth International Conference on Artificial Intelligence in Design (AID), Machine Learning in Design (MLinD) Workshop.

Simon H. A., 1969. *The Sciences of the Artificial*, MIT Press, Cambridge, MA.

Sivaloganathan S. and Shahin T. M. M., 1999. "Design reuse: an overview", Proceedings of the Institution of Mechanical Engineers (I Mech E), Journal of Engineering Manufacture, Proceedings Part B, 213, Number B7, pages 641-654.

Sutherland I. E., 1963. "Sketchpad: A man-machine graphical communication system", Doctoral Thesis, Massachusetts Institute of Technology (MIT), Cambridge, MA. Also available in abridged format from IFIPS Proceedings of the Spring Joint Computer Conference, Volume 23, pages 329-346.

Terry M. and Mynatt E. D., 2002. "Supporting experimentation with side-views", Communications of the ACM, Volume 45, Issue 10, pages 106-108.

Turo D. and Johnson B., 1992. "Improving the visualization of hierarchies with treemaps: Design issues and experimentation", Proceedings of the Third IEEE Conference on Visualization, pages 124-131.

Ullman D. G., 1994. "Issues critical to the development of design history, design rationale and design intent systems", Proceeding of the International Conference on Design Theory and Methodology (DTM), ASME Design Engineering Technical Conferences (DETC), Volume 68, pages 249-258.

Venzin M., von Krogh G., and Roos J., 1998. "Future research into knowledge management", *Knowing in Firms: Understanding, Managing and Measuring Knowledge*, (von Krogh G., Roos J., and Kleine D., Eds.), Sage Publications, London, UK, pages 26-67.

von Krogh G., Roos J., and Kleine D., (Eds.), 1998. *Knowing in Firms: Understanding, Managing and Measuring Knowledge*, Sage Publications, London, UK.

Ware C., Hui D., and Franck G., 1993. "Visualizing object oriented software in three dimensions", Proceedings of the Conference of the Centre for Advanced Studies on Collaborative Research: Software Engineering (CASCON), IBM Centre for Advanced Studies Conference, Volume 1, pages 612 – 620.

Wattenberg M., 1999. "Visualizing the stock market", Proceedings of the ACM Computer Human Interaction (CHI) Conference, Extended Abstracts on Human Factors in Computing Systems, pages 188-189.

Wegner D. M., 1987. "Transactive memory: A contemporary analysis of the group mind", *Theories of Group Behavior*, (Mullen B. and Goethals G. R., Eds.), Springer-Verlag, New York, NY, pages 185-208.

Wenger E., 1998. *Communities of Practice: Learning, Meaning and Identity*, Cambridge University Press, Cambridge, UK.

Wiemer-Hastings P., 1999. "How latent is Latent Semantic Analysis?", Proceedings of the Sixteenth International Joint Conference on Artificial Intelligence (IJCAI), pages 932-937.

Ye Y. and Fischer G., 2002. "Supporting reuse by delivering task-relevant and personalized information", Proceedings of the Twenty-Fourth International Conference on Software Engineering (ICSE), pages 513-523.

Zack M. H., 1999. "Managing codified knowledge", Sloan Management Review, Volume 40, Number 4, pages 45-58.

Zack M. H., 2000. "Researching organizational systems using social network analysis", Proceedings of the 33rd Hawaii International Conference on System Sciences (HICSS), Volume 7, page 7043.

Zieliński K., Laurentowski A., Szymaszek J., and Uszok A., 1995. "A tool for monitoring heterogeneous distributed object applications", Proceedings of the Fifteenth International Conference on Distributed Computing Systems (ICDCS), pages 11-18.

www.ingramcontent.com/pod-product-compliance
Lightning Source LLC
Chambersburg PA
CBHW071148050326
40689CB00011B/2024